Reducing Airlines' Carbon Footprint

Reducing Airlines' CARBON FOOTPRINT

Using the Power of the Aircraft Electric Taxi System
A Mixed Methods, Multi Case Study

DR. THOMAS F. JOHNSON

NEW YORK

LONDON • NASHVILLE • MELBOURNE • VANCOUVER

Reducing Airlines' CARBON FOOTPRINT

Using the Power of the Aircraft Electric Taxi System

Published in New York, New York, by Morgan James Publishing. Morgan James is a trademark of Morgan James, LLC. www.MorganJamesPublishing.com

Proudly distributed by Ingram Publisher Services.

A **FREE** ebook edition is available for you or a friend with the purchase of this print book.

CLEARLY SIGN YOUR NAME ABOVE

Instructions to claim your free ebook edition:
1. Visit MorganJamesBOGO.com
2. Sign your name CLEARLY in the space above
3. Complete the form and submit a photo of this entire page
4. You or your friend can download the ebook to your preferred device

ISBN 9781631950810 case laminate
ISBN 9781636980966 paperback
ISBN 9781631950827 eBook
Library of Congress Control Number:
2020934001

Cover Design by:
Christopher Kirk
www.GFSstudio.com

Interior Design by:
Chris Treccani
www.3dogcreative.net

Morgan James is a proud partner of Habitat for Humanity Peninsula and Greater Williamsburg. Partners in building since 2006.

Get involved today! Visit: www.morgan-james-publishing.com/giving-back

Dedicated in Loving Memory
to Robyn

TABLE OF CONTENTS

LIST OF TABLES

PREFACE

In order to continue expanding the airline industry, more consideration must be given for how to make it more affordable, more sustainable, and more efficient for airlines, employees, and passengers. In *Reducing the Airlines' Carbon Footprint Using the Power of the Aircraft Electric Taxi System*, I highlight one avenue to meet these needs.

This book demonstrates that there are ways to improve airport area air quality and reduce aircraft operating costs, while also improving airport accessibility and increasing ramp safety for the personnel who work at the airports. One way airline managers can reduce aircraft operation costs is to make better use of aircraft and airport facilities through calculated capital investment. Airlines can reduce their carbon footprint, achieve significant cost savings, decrease sound pollution, and benefit the environment by not using the aircraft main engines to taxi. However, the airline industry has been unwilling to adopt ways to taxi aircraft without using thrust from the main engines.

A promising opportunity lies with the aircraft electric taxi system. In the following chapters, I address the features and benefits of the aircraft electric taxi system, as well as some potential drawbacks. As of the publication date, there are no aircraft equipped with an electric taxi system in commercial service. Yet, the electric taxi deserves more careful consideration as it offers the following advantages:

- The electric taxi system can reduce operating costs.
- The electric taxi system changes pushback, eliminating the need for tractors.
- The electric taxi system saves on fuel during taxiing and engine start before take-off.
- The electric taxi system will help with engine maintenance by decreasing run time.
- The electric taxi system improves engine efficiency through lower run time and reduced Foreign Object Debris (FOD) damage.

ETS = Ground Efficiency

The ETS is a revolutionary innovation that will save airlines time and money.

Every flight.

| Reduced time and more dependable departures are key factors for success in competitive markets | Saving ground time on every flight sector is the key to optimizing your crews and resources | Better aircraft utilization leads to optimized network scheduling — more cities, more flights |

In *Reducing the Airlines Carbon Footprint Using the Power of the Aircraft Electric Taxi System*, I outline a way to realizing greater efficiency for aircraft, airlines, and airport facilities.

ACKNOWLEDGMENTS

Airbus Industries
American Airlines
Embry Riddle Aeronautical University
Federal Aviation Administration
Honeywell Aerospace
Institute of Electrical and Electronics Engineers
Kevin Anderson and Associates
Morgan James Publishing
Northcentral University
Robyn Bonfy Johnson
Safran S.A.
Spartan College of Aeronautics
The Boeing Company
United Airlines
University of Phoenix
Wheel Tug PLC.

Honeywell / Safran EGTS Mounted to the main landing gear wheel

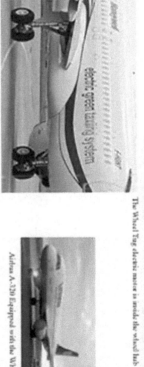

Airbus A-320 Nose Wheel with Wheel Tug Electric Taxi installed

The Honeywell / Safran EGTS demonstrates the caster link turning using the nose wheel

Airbus A-320 Equipped with the Honeywell / Safran EGTS

The Wheel Tug electric motor is inside the wheel hub

Boeing 737 Equipped with the Wheel Tug Electric Taxi

Airbus A-320 Equipped with the Wheel Tug Electric Taxi

CHAPTER 1:

Introduction

Commercial airplanes are getting more expensive to operate for several reasons, including the costs for airport expansions, rising wages of airline employees, and the cost of jet fuel, which are all passed on to the airlines and then to paying customers. In fact, research reveals that jet fuel is 40% – 60% of an airline's operating cost. In order to keep air travel affordable, the airline industry must focus on reducing the costs of aircraft operation. Since everyone flies the same aircraft at the same speed, airlines can distinguish themselves by improving efficiency on the ground. Reconsidering taxiing will improve efficiency in fuel consumption, airport facility access, and improved ground handling operations.

Aircraft use thrust from their main engines for all ground movements on the tarmac. A reduction in airline operational costs could be accomplished in part through new ways to taxi the aircraft. The thrust from the aircraft gas turbine engines for taxiing operates less efficiently than at cruise because they are designed to operate at higher thrust settings. Engine thrust taxiing is inefficient and contributes to high fuel costs, additional engine maintenance, and poor airport area air quality. Some airlines have implemented single engine taxiing to reduce fuel costs and emissions. However, the single engine taxi is subject to greater thrust settings during ground movements and needs more attention from the individual airline operations managers.

This operational inefficiency is not a problem for the Aircraft Electric Taxi System (ETS) which is comprised of motors installed inside the aircraft wheels that propel the aircraft during taxiing and use only electrical power.

Using electric power from a battery and/or a small auxiliary power unit (APU) located in the tail of the aircraft to drive the electric motors would reduce the fuel consumption of the main engines. Additionally, revisiting the way that aircraft taxi may decrease the congestion of aircraft between terminal gates and runways in commercial airport environments.

There is likely to be initial pushback against the electric taxi from airlines since (a) this new mechanism is seen as a major revolution in aircraft ground handling operations; (b) additional training will be required to operate the aircraft under electrical power; (c) the increased ground maneuvering efficiency due to electric taxi system installations could make the non-electric taxi equipped aircraft seem disabled or inefficient; and (d) there is a hypothetical risk that the electric taxi system could create logistical problems resulting in flight delays, cancelling out any value added benefits. However, these potential drawbacks can each be addressed and do not outweigh the potential benefits of the aircraft electric taxi system.

Competition in the international aviation industry has always been a concern for individual airlines. Market pressures have stimulated mergers and acquisitions in the airline industry that were necessary to reduce operational costs and increase profits. Many airlines are unable to cope with marketing strategies that give competing airlines a leading edge. The airline industry is a business that makes investments and expects to get returns on their investments with profits. The business models for most domestic airlines are very similar; however, some airlines require a shorter time for their return on investment. It can be difficult when starting a new investment project increasing operational efficiencies to know the ideal timing to forecast their return on investment. The question of investment and return will contribute to airlines' attitudes towards adopting the aircraft electric taxi system. The following three questions must be addressed. 1. Does the cost of jet fuel influence airline executives' decision in adopting ETS technology? 2. Does the measurable payback period for a return on investment influence airline executives' decision of adopting ETS technology? 3. Does the amount of government financial assistance influence airline executives' decision of adopting ETS technology?

With small profit margins and extreme competition, the risk of not taking advantage of ways to decrease operating costs through reduced fuel consumption and engine maintenance could mean the difference in whether an airline remains in business or folds. So much attention has been given to efficient operation while the aircraft is in flight, and continued research is needed in this area, taking notice of the latest cost saving technology available. Yet, a necessary addition to this conversation must be to look for cost reduction opportunities while the aircraft is on the ground.

CHAPTER 2:

Benefits of the Aircraft Electric Taxi System

The benefits of the ETS are many.

Non-Time Benefits

- Slot Availability
  Reducing terminal, taxiway, and runway time per aircraft could result in preferential slot access

- Slot Creation
  Engines off for taxi could open new slot opportunities or move up early morning departure at curfew airports

- Logistics
  Less reliance on ground equipment at outstations

- Nose Landing Gear Wear
  Eliminate non-linear tractor impact on gear assemblies

- Air Quality
  Lower secondary smog from aircraft could assist with airport relationships and slot availability.

The following chapter reviews aircraft fuel consumption and aircraft time on the ground during airport visits. Then, it examines the cost of ownership and operation of the airline ground support equipment (GSE), including aircraft ground-handling safety and ground operations performance. Airport accessibility, the cost of aircraft engine maintenance due to foreign object damage incurred while the aircraft is on the ground, and technological advances in the aviation industry as related to aircraft ground operations are also discussed. The chapter then considers the future of the single aisle narrow-body aircraft and landing gear durability. Finally, the drawbacks and complications related to an aircraft electric taxi installation and airport area air quality are addressed.

Fuel Consumption Evaluation

The price of jet fuel rises every year, and there is unpredictability in oil producing nations. Additionally, there is a corporate demand to cut costs and a global demand to reduce emissions. These are just a few of the current issues creating the demand for more fuel-efficient aircraft operations.

The amount of fuel consumed by an aircraft during its operation from start-up through to taxi and takeoff, to cruise, to approach for landing and taxiing on arrival is the major variable in cost incurred by all commercial airlines. Many of the factors can be influenced by airlines with proper operations planning and strategies. Operational improvements can be expressed in terms of operational efficiency, which is the combination of ground and airborne efficiency.

Applying vital decisions for new airline routes, aircraft utilization, and aircraft ground operations are important factors for airline decision making. Fuel savings represent the main reason why electric taxiing should be the preferred mode of aircraft ground movement. Savings can be calculated as the difference between taxiing fuel consumption when using the main engines and fuel consumption while using only the auxiliary power unit multiplied by the time the aircraft taxies. To quantify the fuel consumption while the main engines are running, it is essential to determine factors such as the number of stops, the number of turns, and the number of acceleration events during taxiing. There will be differences in the time of operation and the style of piloting during weather conditions at each airport. For the precise estimation of the fuel consumption evaluation (FCE) while the main engines are running the following formula is used:

$$FCE = \sum_{m-1} t\, mi \times f\, mi$$

Where $t\, mi$ is the time spent by the aircraft in state m, $f\, mi$ is the fuel flow while the aircraft is in state m. States m are: stop, accelerating after stop, turning, taxiing at constant speed or braking. The thrust setting varies for these states. It is between 4% and 9% thrust setting for a standard jet aircraft. Furthermore, according to flight crew interviews, the upper values may differ more based on the nature of the flight crew.

It may be necessary for individual airlines to make their own calculations about their crew behavior for precise estimations. For less precise estimates, the author used the model detailed in the base of aircraft data maintained by EUROCONTROL. The average ground operations thrust setting used for the model is 7%. Some operators say this is overestimated in most cases.

Table 1

Aircraft Engine Fuel Consumption at Taxi Speed
Engine Type Fuel Flow with 7% thrust (kg/s).
CFM56-5B6/3 0.095
CFM56-581/3 0.109
CFM56-5-A1 0.101
CFM56-586/P 0.097
PW6122A0, 109 0.109
PW6124A 0.114
CFM56-7822/3 0.099
Auxiliary Power Unit (APU) 0.034 (max value)
GTCP 131-9A.
Data from Aircraft Emissions Databank
(http://easa.europa.eu/environment/edb/aircraft).

Some studies conducted by industry experts state that as the cost of aviation fuel rises each year, there should be fuel saving initiatives sponsored by the government and/or aviation community. The International Air Transport Association started an initiative to help airlines deal with the impact of increasing jet fuel prices and has encouraged airlines to increase their operating effectiveness by opening routes that are more direct, re-scoping routes that are ineffective and improving traffic flow wherever possible.

Another method developed to reduce fuel consumption uses the continuous descent approach. As the aircraft approaches the airport, the pilot usually descends in a stair-step fashion reducing engine power and waiting for the controller to clear the flight down to the next level. The continuous descent approach consists of a constant angle as the aircraft descends to landing, which consumes less fuel than the stair-step descent.

Other studies detail aircraft flight path optimization and how it can be designed to minimize environmental impacts of aircraft around airports during approaches. The main objective is to develop a model of optimal flight paths considering jet noise, fuel consumption, constraints, and extreme operational limits of the aircraft while on approach to landing. One industry study modeled the optimal control problem using a two-segment approach. The first is an optimal trajectory during approach to the runway of about 1060 ft. /mn. The second is the aircraft alignment with the runway. In the model, noise and emissions are reduced by -4dB and fuel consumption is reduced by -10% to -20%. This model can be integrated into the aircraft flight management system and the autopilot.

Environment & Safety

Engines off in the gate area for push out, taxiing, stop-and-go queue and after landing

- Greatly reduces ground noise, improving the environment for both passengers & neighbours

- Work environment for ground personnel is quieter and safer

- Reduces emissions

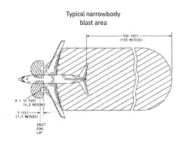

While the International Airline Transportation Association does research on ways for the airlines to reduce their cost of operations in the air, the International Airline Transportation Association has not published information related to fuel savings during aircraft ground operations. The need to reduce aircraft fuel consumption while taxiing will have a major impact on aircraft operating costs. For example, there is potential to increase efficiency with the new pushback option ETS offers.

Time Savings: Pushback

- Average time from pushback to taxi forward: 5 minutes 35 seconds

- In 2% of cases, this rises to 13 minutes

- Airlines pad schedules to allow for these delays

- This is lost time that the ETS helps reclaim

Source for 6,723 pushbacks:

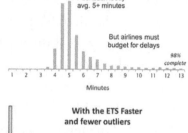

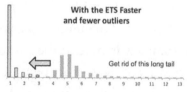

Pushback savings will be direct savings by eliminating the need for a pushback tractor. Depending on the given airport handling cost, direct savings seem to be between $200 and $300 per pushback. Some studies claim $335 per pushback which represents the cost of fuel, maintenance, but not depreciation of a given pushback tractor per one pushback operation. If a nose-in, nose-out stand is used, there will be no pushback savings. How much the change would decrease insurance costs is questionable since there may be a rise in the number of accidents incurred on ground because of flight crews' inexperience with backwards moving operations during the first years of electric taxi operations.

Just as with all airline procedures, there will be a need for operator training processes as electric taxiing is established. Moving backwards with the electric taxi systems could be difficult for pilots because they do not have any rear or side views; however, cameras can be mounted on the landing gear with a display screen for the cockpit to give the pilot a full view of the aircraft's position, and ground personal, who are currently used during tug operations, could still be used for this scenario.

Even if aircraft are equipped with a parking assistant system, it will still be the responsibility of the employees on the ground to prevent collisions. The training of ground personnel will increase the cost of ground operations in the first years of the project because of a higher insurance cost and time spent in training. This means that the savings from the electric pushback will be less significant.

It is possible that some airports will not invest in the ground personnel's training for operating the electric taxi, and air operators will have to use pushback tractors during the first few years. Nevertheless, using the electric taxiing system will simplify gate operations and the time of pushback will be shorter at most airports. Aircraft that are equipped with ETS will also have better maneuverability, which will enhance the use of the tarmac area and lead to greater airport capacity. These savings will induce the aviation industry to support electric taxiing systems, and a scheme to reduce airport usage fees for electric taxiing systems may be applied to encourage the establishment of electric taxiing.

Streamlined Operations

Conventional Pushback

Pilot
Tug Driver
Wingwalkers
Ramp Supervis.
Controller
Pushback Tractor
Towbar
Communications Hardware
Jet Blast Safety Margins
Collision Safety Margins
Pushback Clearance
Pushback
Communications Link Disconnection
Pin Removal
Ground Crew Clearance
Engine Start Procedures
Taxi Clearance

Personnel
Ground Equipment
Safety Factors
Processes

ETS Equipped Pushback

Pilot
Ramp Supervis.
Controller
Collision Safety Margins
Reverse & Taxi Clearance
Reverse Operation

The aviation industry has been focusing on hotter running and more efficient gas turbine engines to reduce the amount of aircraft fuel consumption. The hotter the gas turbine engine can operate, the more efficiently fuel is used and the fewer damaging air pollutants are emitted. As a result, aerospace engineers are pressured to come up with new ways to qualify component parts that can survive the harsh temperatures (over 2000° F) and high dynamic loads of the gas turbine engine environments, which facilitate increased fuel efficiency. As the normal operating temperature of the gas turbine engine (GTE) is increased, fuel efficiency can be increased.

When a gas turbine engine runs at a higher temperature and pressure, it extracts more energy from the fuel allowing the engine to run more efficiently; however, extreme thermal conditions and extensive running speeds increase the stress on the turbine blades and rotors. These conditions demand a new class of materials. Gas turbine engine components are subject to degradation and stress through thermal interaction, erosion, creep, hot corrosion, and vibration. This increased amount of stress on the engine components causes premature failures, shortening the service life of the engine.

Thankfully, there are other ways to reduce aircraft fuel consumption without redesigning engines to withstand the additional stress from higher operating temperatures. The airline industry could focus their attention on ways to reduce fuel consumption during ground operations rather than the technologically difficult approach of focusing on turbine efficiency. For example, if an aircraft could taxi without the use of the main engines, the airline could save approximately 4% in annual fuel consumption.

The recent increase in jet fuel prices continues to have an impact on an airline's cost of operations throughout the world. This has resulted in bankruptcies and a significant reduction in flight operations and services among airlines throughout the industry. The jet fuel and flight crew are the two major factors of operating cost. Just a one percent increase in jet fuel prices translates to a $530 million increase in cost for the United States airline industry.

Recent airline industry reports from the International Airline Transportation Association have stated that the airline industry could be up against a challenging future of unforeseeable financial losses, following the last 60 months of increasing oil prices. The latest figures from the International Airline Transportation Association show that if oil prices stay high, the airline financial losses could exceed $6 billion dollars. When an airline uses an alternate means to taxi aircraft, the real time operational costs can be reduced through less fuel burned.

One airline simulated a model that used the towbarless towing vehicle to move aircraft to and from the passenger gate area. This simulation model also addressed aircraft relocation from the maintenance hangar to the terminal hub. The study researched the opportunity to decrease costs through the reduction of jet fuel consumption by using the towbarless towing vehicle, and to identify if the airline's high purchasing costs are justified. In addition, the study used the simulation to analyze the annual fuel savings, as well as the justification of operating costs for these towbarless towing vehicles. The results were promising and enabled the airline to clearly evaluate the airline's costs and benefits for purchasing new towbarless towing vehicles.

The less fuel burned while the aircraft is taxiing, the greater the overall operational savings. A group of aerospace engineers considered the departure procedures at London Heathrow airport with an eye to reducing fuel consumption. These improvements could decrease the average delay for aircraft waiting take-off, reducing the amount of jet fuel consumed and leading to lower operating costs and improved environmental benefits. The engineers addressed two potential problems at various stages of the aircraft departure system. Then, two solution systems were evaluated and proposed. If the overall amount of fuel used during taxi is reduced, the cost to taxi an aircraft will be reduced. The authors presented the first system and the opportunity to improve the aircraft take-off sequencing by providing a decision backup system to the controller who schedules the aircraft for takeoff.

Other airlines are increasing the seating capacity in some aircraft, increasing the number of passenger seats flown by their smaller aircraft. For instance, *The Wall Street Journal* notes that Delta Airlines and United Airlines have taken steps such as using smaller planes on routes in addition to taking aircraft out of their fleets to reduce the number of flights per day and increase passenger loading. These actions have been in the interest of reducing fuel consumption calculations and increasing passenger seat per mile calculations.

A typical aircraft may taxi on to and from runways with thrust force developed by its engines. A significant amount of fuel is burned by the engines during a typical aircraft taxi profile before and after each flight. In many cases, the main engines provide more motive force than is required to complete a successful taxi profile. In that regard, engine-thrust taxiing is inefficient and contributes to high fuel costs and ground level emissions.

Airlines have to increase technological efficiencies by replacing old fleets with advanced, fuel efficient aircraft to cope with the impact of higher jet fuel cost volatility and stay profitable. Major engine manufacturers responded to that pressure. Both Pratt & Whitney's PW1100G geared fan engine and the LEAP 1A engine from CFM

International promise a reduction of about 15% – 18% in fuel consumption.

The reception of these engines by the airlines has been so positive that airframe manufacturers Airbus and Boeing plan to offer essentially new versions of existing single-aisle, narrow-body aircraft that can accept the high-efficiency engines. The worldwide production of gas turbines includes the commercial and military aviation markets, as well as non-aviation markets for electrical generation, marine applications, and mechanical power. Analysis shows that the value of production of gas turbines worldwide was $82.5 billion for 2018, up from $77.9 billion in 2017.

Aircraft gas turbine engines are designed for optimum performance at aircraft cruise or higher power settings. Recent studies have found that the operation of the main aircraft power plants at the low powers for ground operation is very inefficient and therefore very wasteful of fuel. In general, the actual aircraft performance can be determined by how the aircraft is operated subject to operational constraints. The efficient operational procedures are those in which the actual fuel burn falls close to the theoretical minimum. Also, the operational factors to reduce the fuel consumption per passenger mile include the increasing load factor, optimizing the aircraft speed and fuel weight, limiting the use of auxiliary power, eliminating non-essential weight, and reducing taxiing.

Volatility in fuel prices and its impact on air carrier firms has drawn interest from the air transport industry and financial markets. Per the International Airline Transportation Association, the global airline industry's fuel cost was estimated to be $207 billion in 2018, constituting 33% of operating expenses at $110/barrel of Brent oil. This is an increase of $31 billion over 2017 and is almost five times the $44 billion fuel expenses in 2003.

The cost of jet fuel is the second largest expense to operate an airline. When airlines face near-bankruptcies they react by reducing flights, equipment, personnel, and services. Airline travel is also impacted by economic downturns and security threats or incidents. Furthermore, there is a new-generation airline model that is based largely on shorter distances and smaller airplanes that require less fuel to operate. This allows airlines to service more compact areas in a more efficient way. Airfares generally increase in recessions and severe downturns. The first decade of the twenty-first century is characterized by two severe recessions amounting to at least 7 of the 10 years between 2000 and the end of 2009. This means airfares are difficult to raise and various means of disguised revenue rising are utilized. These include higher luggage charges and elimination of services related to on-board passenger needs. In addition, free meals and snacks have been discontinued and most in-flight food services are subject to a charge. Addressing rising costs is necessary to keeping passengers happy and airlines in operation.

Ground Taxi Time Evaluation

Using the electric taxi system can save time during ground operations, especially when an airport has an inadequate number of pushback tractors for the number of daily aircraft movements. This represents a problem during peak hours as waiting for a tractor may cause a delay. Other time savings may be achieved when special ground maneuvers are necessary. An electrically driven undercarriage may better perform special ground maneuvers than an engine, reducing pushback related operations since there is no need to use tow bars.

Additionally, for the aircraft equipped with the electric taxi system, more time can be saved during de-icing operations, as the de-icing trucks can get closer to the aircraft since its engines are not running. Such a measure may lead to better use of the hold over time. These time savings are in minutes and are, admittedly, minor without any significant impact on the aircraft circulation planning process. However, they do contribute to passenger satisfaction with the airline.

While these minutes may seem minor, industry experts believe that the time saving will increase for the electric taxi users. Adding together the minutes saved from several flights can create a time slot for another flight. Realizing this possibility requires the flexibility of airline operations.

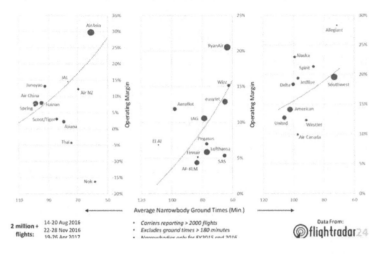

One factor that will need further consideration is the taxi time. Since the electric taxiing system will not allow for higher speeds, the taxi time may actually be longer than the taxi time using main engines. So, time savings gained from faster gate operations may be reduced by taxi times. This review will not include any savings or costs related to taxi time changes.

Airline turnaround time is defined as the time required to unload an airplane after its arrival at the gate and to prepare it for departure again. The previous section identified ways to address the rising fuel prices. Since there is very little control over fuel costs, however, one of the ways airline companies can cut costs without compromising quality is by reducing the aircraft turnaround time. In order to cut costs and increase performance, airlines work constantly on reducing turnaround times as these smaller and larger adaptations are important process innovations. They are crucial for competitiveness and a key parameter in determining the profitability of an airline company. The airline turnaround times are measured, recorded, and are of public record. They have a direct impact on pre scheduled gate arrival and departure times. In times of stiff airline competition, the airlines will boast about their best on time departure record to get more passengers.

When airplanes are full of passengers, their gas mileage and passenger seat per mile is better. The smaller more fuel-efficient airplanes require shorter runways and less costly outdoor passenger loading equipment to plane and deplane. Smaller planes mean that carry-on luggage is often gate-checked, then dropped at plane entry, and retrieved on plane exit. This has implications for how people board and depart the plane. The above requires revenues to respond to new conditions within the airline industry, which can be a challenge as most U.S. domestic airlines operate on the verge of bankruptcy.

With the electric taxi system, there could be some fuel savings due to the lower weight of aircraft during flight, as less fuel needs to be carried because less is needed for taxiing in. However, such savings are theoretical since it is probable that regulation will require surplus fuel be carried on board the aircraft for taxi in case the electric taxi is inoperable. An additional factor that must be considered is taxi planning, which studies aircraft routing and scheduling on airport tarmac. This is a dynamic plan which must be updated almost every time a new aircraft enters or exits the system. Sky and ground operations can get crowded in areas of heavily congested airports. The national air traffic control system needs serious upgrading, as plane passage from area to area is not always smooth. There is currently a lack of information on this topic, so further study is needed on how improving air traffic control can improve airport and airline efficiency and savings.

Ownership and Operation of Airline Ground Support Equipment

The groundside of airport operations involves anything outside the security gates. The airside of airport operations involves anything on the inside of the security gates, not accessible to the public without going through security screening. Airport ground support equipment includes carts for moving materials and people, tugs for airplane pushback, forklifts and lifts, air conditioning tugs, containers, belt loaders, other specialized equipment, and vehicles that provide power to the aircraft. Airport ground support equipment can encompass all types of vehicles. Many are light duty trucks used for fueling on the airport tarmac and airplane maintenance on the airside of airport operations. This equipment is usually part of the airport airside operations.

It is possible for the airport ground support equipment to be owned by either the airport or the airline operator; however, most domestic airlines maintain or own the ground support equipment they use. Some airlines might have a full service leasing agreement or pay an hourly rate for the equipment only when they use it. The buildings and stationary infrastructure are usually owned by the airport and leased to airline companies. The airport property management staff usually has responsibility for the cost of improvements, electrical equipment, and vehicle network infrastructure of airport owned equipment

There are areas for revenue growth in airport management. For example, airports that purchase and own the ground support equipment charge airline companies a rental fee each time they use the equipment. Airports gain more revenue by providing the ground services on the tarmac. Aircraft tugs and pushback tractors tow the aircraft into airport areas where the aircraft cannot start their main engines. These tugs and tractors are used generally in the areas between the terminal and the maintenance base as well as between the taxiway and the terminal.

Ground support equipment is expensive whether purchased new or used. Many of the replacement parts needed for the equipment are specialized and can only be procured from the equipment manufacturer. This monopoly on maintenance parts makes the airline ground support equipment expensive to own. There are also labor rates to consider. Good ground support equipment technicians are in short supply. Many companies do not have enough technicians on staff due to a shortage of certified mechanics. This means there may be a backlog of broken equipment as mechanics might be overworked.

Yet another issue with ground support equipment is the potential for accidents and delays. Industry data indicates airport ramp accidents cost major airlines at least $10 billion a year globally. These ramp accidents occur on the airside of airport operations. Airport operations are affected by these accidents that result in damaged aircraft, personnel injuries, damaged facilities, and damaged ground support equipment. Greater safety measures could be sought by the airline operational managers to begin addressing these issues.

The International Airline Transportation Association foundation initiated one means of addressing these accidents when it launched the Ground Accident Prevention program in 2003. This program develops products and information in a user-friendly format using e-tools designed to eliminate incidents and accidents on airport tarmacs and taxiways as well as during aircraft movement into and out of hangars. Alternate ways to move aircraft around the tarmac could increase the margin of safety.

Some of the issues and equipment listed above could be eliminated with the application of the aircraft electric taxi system. For example, the cost of the pushback tug and personnel to operate it can be itemized and deducted when calculating the aircraft electric taxi system installation payback analysis.

Ground Operations Performance

The electric taxi system components can be electronically configured to work as a regenerative braking system as seen on most modern day golf carts. A regenerative brake is a mechanism in which the electric motor that drives the vehicle operates in reverse during braking. It is a system of energy recovery that reduces the speed of the vehicle. It is done by transforming the kinetic energy into a form that can be used instantly or stored for later use. The components of regenerative brake systems include an actuator and a storage device. These are required to capture and store energy.

In an electric taxi system, the actuator is the electric machine. The energy storage space is the battery, as it is the electrical power source. Regenerative braking is the act of turning a motor into a generator. In this way, it slows down the load that it was driving. Regenerative braking puts a little of the energy back into the traction pack and would put the absorbed energy back into the aircraft batteries. It saves the aircraft brakes from being damaged and gives a boost to the aircraft battery every time the aircraft stops. With the installation of the aircraft electric taxi system the need for the anti-skid braking system can be eliminated.

Currently, aircraft employ an anti-skid brake system to aid in safety and mobility while on the ground. However, the necessity of the anti-skid braking system can be evaluated after the first few aircraft are equipped with the aircraft electric taxi system. Aircraft with the electric taxi system should rely less on the anti-skid braking system after the aircraft lands and comes to a full stop. The aircraft functions of taxiing and steering normally assisted by the anti-skid braking system can be handled entirely by the electric taxi system. This could be the focus of a continuing study.

The anti-skid braking system has been important in aircraft safety by addressing speed and traction. Speed reduction without the anti-skid braking system installed is nonlinear, time varying, and uncertain during the braking process after the aircraft lands on the runway. Overrun incidents in past years occurred mostly on wet runways. Therefore, the confirmation of wet runway or dry runway status is crucial, particularly for civil aircraft to optimize the greatest aircraft tire to pavement friction and to be certain about the safety of aircraft braking. Engineers have found that the robustness, adaptability, and anti-interference ability of the runway status can be improved with the use of the anti-skid braking system by using simulation and dynamometer testing.

The reliability and safety granted by the anti-skid braking system can largely be assumed by the electric taxi system, as the electric taxi system can handle most features covered by the anti-skid braking system when the electric taxi system is electronically configured with the regenerative braking feature. The electric motors mounted in the aircraft wheels will power the aircraft off the runway and steer it to the airport terminal, all controlled by the pilot with the aircraft main engines shut down.

Aircraft equipped with the electric taxi system will give airport designers the ability to utilize surrounding areas that in the past could not accommodate jet aircraft using main engine thrust for taxi power. As long as there is enough clearance for the aircraft wings and a smooth ground surface below, an aircraft equipped with the electric taxi system could taxi anywhere. This newly found application of aircraft mobility will open up new airport design standards.

Airport designers have stressed the importance of standardization within the global aviation community to maintain a common airport infrastructure. This infrastructure could involve common accessibility to airport taxiways, aircraft service facilities, and passenger terminals. Aircraft taxiways could have the same standardization for pilots traversing within an airport as the aviation airways that pilots use for navigation from airport to airport are standardized. Airports should standardize based on the best ideas available, enabling the best use of resources. Time, space, materials, manpower, and cooperation between participating governments on a voluntary basis are important aspects in realizing this global standardizing. Airport and aircraft accessibility planning are important components of that process. Industry experts have examined different aspects, with a focus on retaining the best intellectual tools, lessons learned, and new technologies.

Also important are the multimodal metro area planning, acceleration of successful regional trade agreements, and examination of the influence of policy and regulation within the local government. The airport infrastructure should take on a different design as the electronic taxi system is adopted by airlines.

The team of air traffic controllers at London Heathrow airport in the United Kingdom presented the development of a management system for airport terminal traffic flow using an intelligent optimization method. This system allows air traffic controllers to assign runways to cope with the unbalanced traffic flow from and/or toward different directions and computes the optimal arrival or departure time for each flight. This is a complex ground traffic advisory system to assist airport managers in orchestrating complex terminal traffic in a more efficient manner.

This system should ultimately minimize the overall flight delay in the entire airport and maintain a high level of safety at the same time. Heathrow uses multiple objectives pertaining to overall airport system delay, throughput, maximum individual delay, and runway balance. This system is suitable for airports with multiple runways. Furthermore, the simulations are based on the real traffic mix for four of the 30 busiest airports in the United States, and the results of the simulation prove the feasibility of the system.

Another study focused on yet another aspect that influences airport design: airport ground collisions and the serious implications from both a safety and a commercial point of view. According to the Flight Safety Foundation, the aircraft industry loses up to $11 billion worldwide each year due to accidents and incidents on the ground. Furthermore, they found that many incidents resulting in human injury and aircraft damage are related to aircraft maneuvers on the ramp and in taxiway areas. In addition, near misses frequently occur and are rarely formally reported. An analysis of air safety reports from NASA's Aviation Safety Reporting System indicates that accidents are likely to occur 43% at the gate stop area, 39% at the gate entry and exit area, and 18% outside the ramp entry area.

The size of a typical commercial aircraft continues to grow, as aircraft designers make aircraft larger to accommodate more passengers and the aircraft engine manufactures make the engines bigger to accommodate the increase in size. The wingspan of commercial aircraft is getting longer. For example, the wingspan of the latest Boeing 737 is 138 feet. The wingspan of the new Airbus A380 is 262 feet. This increase adds to the complexity of aircraft ground maneuvering considering the task of judging clearances or distances between objects. This judgment is completed by the pilot without any aids. The issue of wingspan is such that Boeing is planning on having a folding wing for the 777x so that it will fit into existing gates and permit added fuel efficiency in flight.

Ground operations at airports servicing aircraft equipped with the electric taxi system could take on a new dimension. The electric taxi system eases the backing of the airliner from the passenger gate under electric power. The aircraft can relocate to get out of the way, so the next aircraft can unload passengers at the same gate, then back out of the way to get ready for the next aircraft, increasing single gate loading efficiency. The increased safety of moving the aircraft without the main engines running will add value through the human factors advantage. Human factors in aviation refers to environmental, organizational, and job factors, including human and individual characteristics, which influence behavior at work. In this case, it would be the aircraft gate area, which can affect health and safety of ground support personnel.

While efficient aircraft gate scheduling is important, airport runway scheduling is just as important. The air traffic controllers at Los Angeles International airport performed a study that focused on runway scheduling. The main performance element that is measured with runway management is the separation between aircraft taking off and aircraft landing. The dependency of separation on the leading and trailing aircraft type make the sequencing and scheduling of landings and take offs an important consideration. Careful scheduling and sequencing can reduce the number of long separation times opening up opportunities for new takeoff slots or landings.

Some Federal Aviation Administration air traffic controllers at Atlanta Hartsfield International airport analyzed the aircraft landing problem and the aircraft take-off problem. Their analysis determines the sequence of aircraft taking off from or landing on the available runways at airports in order to optimize given objectives, subject to a variety of operational constraints. Optimally sequencing the takeoff and landing, aircraft may increase runway capacity, although it may not always be possible to implement the solutions. The challenge is to put theory into practice, which involves handling efficiency, safety, competitiveness, robustness, and environmental issues. Furthermore, some suggest that one aircraft may be given precedence over others due to airline preferences, overtaking constraints, or high priority flights.

Airports with multiple runways may operate the runways in either a mixed mode or a segregated mode. In the runway segregated-mode, the runway is solely used for either landing or take-off of the aircraft, while mixed-mode allows both landing and take-off on the same runway. The mixed mode is usually more efficient than segregated mode since alternating landings and take offs on the same runway is effective in reducing or eliminating delays due to wake vortex constraints.

Aircraft equipped with the electric taxi system will help to eliminate some of these problems discussed. Efficiency is increased when the airline ground support equipment and related personnel are not required for aircraft ground movement without the main engines running. Safety is increased as well, and the aircraft can taxi without using main engine thrust. Furthermore, the airport area accessibility by the aircraft will be increased, utilizing the facilities to their fullest advantage.

Airport Accessibility

The electric taxi system will give aircraft better mobility on the ground and better access to airport facilities. Redesign of terminal buildings or redesign of the ramp area can be costly, so ramp redesign is usually not an option. It is in the best interest for a terminal operator or airport manager to use the gates available to the best possible advantage. The electric taxi system will allow aircraft to maneuver around airport terminals without using the main engines and therefore use existing airport terminal gates and facilities to their greatest advantage.

An airport of the future will have the option to be designed for electric taxi system equipped aircraft. This airport could have facilities to accommodate aircraft that can taxi forward, backwards, and even sideways, enabled by the caster action of the landing gear and the new aircraft electric taxi system. As the system gradually gains recognition, airport designers will have the option to find ways to increase aircraft mobility to make better use of airport facilities.

Airport Movement Benefits

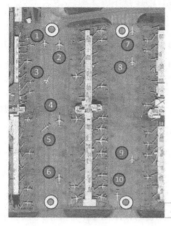

Reduced jet blast allows faster taxi-back clearances.

The ETS equipped aircraft cuts safety clearance delays, increases aircraft movements.

Hubs are congested – even on good days.

A single pushback delay in any of these areas can result in delays for a dozen aircraft or more.

Even a relatively normal pushback can significantly slow operations.

Atlanta

Airline ground operation efficiencies resulting in reduced fuel consumption can be gained through precision aircraft taxi operation and using the most direct routes on the tarmac. The risk of collision with another aircraft or stationary object can also be reduced. At the Seattle Sea Tac airport there are aircraft ground support tracking systems comprised of various proximity detectors. These detectors communicate with each other and the aircraft while on the ground. By comparing the aircraft's position on the ground to other aircraft or stationary objects, the proximity detector activates an acoustic alarm when it senses an object in the path of the taxiing aircraft. This invention to determine the risk of collision on the ground between an aircraft and another object could be integrated into the electric taxi system with the anti-collision alarm for increased mobility, quicker operator response time, safety, and efficiency. The combined technology provides important, real-time information that can be used by airport ground scheduling personnel.

While effective airport ground scheduling is important, effective aircraft arrival predictions are important as well. Aircraft arriving at a gate that is still occupied by an aircraft that has not yet departed is a complete waste of time for the incoming aircraft, personnel, passengers, and airline.

A group of air traffic controllers reviewed a scheduling system developed to effectively orchestrate aircraft arrival into congested airports from a long way out. This method presents an aircraft arrival schedule to the airport as a decision support system. The aircraft arrival scheduling system is comprised of a distributed network of independent aircraft schedulers with a common denominator of sharing airway capacity information. These schedulers help to populate the airways in the skies around the airport as well as the terminal gates at the destination airport. This sharing of information insulates the schedules from unpredictability in estimating long-distance arrival times and allows flexible schedules for short-term arrivals. Precise ground scheduling for passenger aircraft at the departure and arrival airports can increase efficiency, saving time and money.

The scheduling of the terminal gate aircraft population at the destination airport will increase schedule reliability and decrease fuel consumed during ground taxi with the addition of the electric taxi system equipped aircraft. The air traffic controllers can use aircraft descent rate profiles to effectively predict the aircraft arrival at the destination airport. The higher and faster the aircraft is flying, the further away from the field the descent must be started. Descents can easily begin more than 100 miles (160 km) from the field, so it must be ascertained in advance that the runway is available. The descent rate profile mechanism allows feasible long-range arrival schedules to be constructed as well as reliable short-range schedules to be constructed, eliminating scheduling problems while meeting the operational limitations of airline operations.

When aircraft are routed effectively they will have their predetermined gate available at the terminal as soon as they land and taxi. The captain will dock the aircraft at the gate and flight attendants should open the cabin door within 45 seconds after the aircraft comes to a complete stop. At this time, the flight is considered completed. Any unpredicted delays after touchdown could affect the overall flight schedule in a negative way. Aircraft equipped with the electric taxi system that taxi with the main engines shut down will have additional gate accessibility options that could minimize, if not eliminate, unpredicted delays.

A group of air traffic controllers at London Gatwick airport looked at the application of airline traffic control, using departure scheduling and arrival simulation. This can be useful as a flow control tool for airline flight operations and related performance modeling. The study used the primary performance modeling in a 40-second time interval ratio that averages the time delayed getting to the gate and early flight arrival at the gate. Simulation modeling can be used to compare the airline arrival and departure performance from two alternative choices, in addition to experimenting with alternative designs.

The air transport system consists of three major components: air traffic management, airline operations, and airside airport ground operations. Air traffic management begins when the aircraft pushes back from the terminal gate and ends upon arrival at the destination airport terminal gate. Two examples of air traffic management tasks are to ensure that there is enough space in the gate areas between aircraft for safety and to decide which runway the subject aircraft will take off from. Airline ground operations range from aircraft scheduling and crew assignment to maintenance planning.

Airport operations include all processes occurring at and around the airport, including which gate the subject flight will be connected to or which fueling vehicle will serve a certain aircraft. Airport operations involve the management of many flows. Passengers with luggage and cargo are referred to as value flows since those are the flows that generate value to the air traffic system. These can be divided into minor support flows such as catering, cleaning, fuel, de-icing, baggage handling, water, lavatory service and major support flows such as the aircraft and crew. Most of these value flows, except aircraft de-icing, take place at the terminal gate.

In today's typical airport environment, an airplane arrives at a terminal gate, unloads passengers, takes on new supplies and crew when needed, and stays at the gate until the next scheduled flight out. Holding an aircraft at the gate when passengers do not occupy it or it is being serviced may limit access to the gate for other aircraft waiting to unload passengers. As problems arise with scheduling connecting flights and so on, these delays can cause a domino effect disrupting the futures of many flight schedules. Due to the mobility of the aircraft electric taxi system the aircraft can arrive at the terminal, unload passengers, and then taxi away to a remote location for servicing, freeing up the gate for other aircraft to board or disembark passengers. The result is an increase in airport gate utilization and more effective passenger management.

One Plane – Many Gates

Because of tow tractor and jet blast safety issues, one aircraft pushing back and taxiing forward can block many other gates.

Frankfurt, A & B concourses

LaGuardia, Terminal C

The flexibility offered by the aircraft electric taxi system has the potential to save airlines time and money when it comes to passengers and regulations. The Federal Aviation Administration has adopted a new ruling that states airlines cannot keep passengers on board the aircraft while it is on the ground for more than two hours. If the airline defies the rule, it could be fined up to $27,000 per passenger. It would take only a few of these incidents to make aircraft electric taxi system installations part of a corrective action plan, when it come to a route cause corrective action (RCCA) investigation.

Additionally, the electric taxi system equipped aircraft can increase gate scheduling reliability. The assignment of flights to terminal ramp positions is gate scheduling and one of the key activities among operations at the airport. The complication of gate scheduling has increased considerably with the increased density of civil air traffic and the equivalent growth of airports over the past years. Hundreds of aircraft arrivals and departures are handled per day at large international airports. These tasks can increase in complexity if there are unpredicted changes to the related flight schedule on the day of operations due to factors such as delays and aircraft changes.

The addition of the electric taxi system can increase flight schedule reliability. The main components for gate scheduling are a flight schedule with flight departure and flight arrival times and additional detailed flight information focusing on the links between successive flights served by the same aircraft. The main components for airport gate scheduling are the number of passengers and type of aircraft, the destination or origin of a flight, cargo volume, and whether the flight is international or domestic.

This information is important for predicting aircraft timing as it directly affects how efficiently the aircraft can be turned around for its next flight. Turnaround time will be included in the master schedule input data that is vital to effective terminal gate scheduling. Terminal gate scheduling is only as reliable as the input data.

Input data uncertainty in gate scheduling may have several causes: errors made by staff, emergency flights, light earliness or tardiness, severe weather conditions, and flight or gate breakdown. For example, a delayed arrival of one aircraft might cause a chain reaction of late arrivals for other aircraft assigned to the same gate.

Increased efficiency at the gate resulting in fewer schedule delays will be possible as the aircraft electric taxi system technology is applied to the airline operation. The stationary position at the passenger gate after the passengers are unloaded should not be required until the next set of passengers is ready to board the aircraft. In addition, the aircraft can taxi away from the terminal to remote locations for de-icing, refueling, and anything else needed before the next flight. Fewer schedule delays increases business for the airline resulting in bigger bottom line revenue.

Air travel is the most efficient mode of transportation once the aircraft is in the air. It is the airports that can cause flight schedule delays when ground operations are not used to the fullest advantage. The processes and flows at airports form a complex system. Many of the processes are critical to the overall performance of the air traffic control system. The efficiency of the air transportation system is often a function of the individual efficiency of all parties involved.

A further step airports must take is to maximize overall efficiency by utilizing the information made available by collaborative decision-making. The concept of collaborative decision-making resolves issues from lacks in communication with the objective of better predictability and punctuality. When airport logistics refer to the control and planning of all information and resources, they create value for the customers who use the airport. The adoption of the aircraft electric taxi system and the additional aircraft mobility it provides will greatly improve the control and planning of the airport resources.

The aircraft turn around process begins when an aircraft lands on the runway and ends when the aircraft is airborne after runway take off. During turn around many supporting flows are present: baggage and passengers have to be unloaded; the aircraft has to be fueled and cleaned; the lavatory holding tanks have to be emptied; and the supplies of food must be re-stocked. Aircraft equipped with the electric taxi system will reduce aircraft turnaround times and increase airport productivity. A loss of productivity could greatly affect a privately-owned airport's ability to generate revenue. However, the government has pressured privately-owned airports that serve the public to improve productivity and be more financially self-sufficient. Properly operated runway configuration management enables effective aircraft ground traffic control.

The air traffic controllers and airport management at Phoenix Sky Harbor airport in Phoenix, Arizona identified runway configuration management as the focus for research. The outcome of the research was to generate an approach for scheduling arrivals and departures that would identify active runways, runways accepting departures, runways accepting arrivals, or both. The study concluded that some problems are controls of gate sequencing and ground movement. The purpose of effective gate sequencing and ground movement is to increase throughput by effectively managing airport ground traffic, while meeting scheduled aircraft departure and arrival time slots.

The ease of aircraft ground movement on the airport tarmac can be restricted by the airport shape and size. Maneuvering aircraft on the ground, around obstacles, and around each other is normally possible but may have many constraints. Assigned gate location is also a consideration. The ground traffic planner must identify the type of aircraft compatible with specific gate locations within the constraints of the different airline relationships and affiliations. Some components of these problems include flight arrival and departure schedules, airport characteristics and runways in use. Aircraft equipped with the electric taxi system add a new dimension to airport area ground traffic management.

Overcoming Airport Constraints

HOURLY PASSENGERS/GATE

Comparison of airports with
40-50 million passenger
movements in 2017.
(GRU – 38 million)

The ETS comes at no
cost to airports, reduces
traffic and ground clutter,
and makes airports safer
and quieter.

HOURLY PASSENGERS/RUNWAY

The key to effective airport area ground traffic management is to effectively get the aircraft and the passengers in the right place at the right time. Keeping the passengers properly entertained before and after each flight can have a direct impact on the airport revenue stream. Departing, arriving, and transfer passengers at the airport have different needs. Each of these types of customers will use different shops and service centers within the airport facility. A good example might be that transfer passengers have a set amount of time on their hands as they wait for their connecting flight. Transfer passengers are more likely to add value to local vendors in their time at the airport than the other types of passengers who are transient and may have less time to shop before or after a flight. Aircraft equipped with the electric taxi system could taxi into areas with the main engines shut down that would normally not be accessible with the main engines running to get passengers closer to shopping and entertainment, making better use of their time spent at the airport.

There are many large global airports that serve as airline hubs for transferring customers. The operations management of these airports has recognized the value of transient passengers and the importance of catering to transfer passengers and ensuring their comfort while they are waiting for their next flight. Understanding the specific needs of transfer passengers is very important to maintain growth in today's competitive airport environment.

Transfer passengers behave differently than arriving and departing passengers. Research shows the effect of service quality on transfer passengers at the Inchon International Airport. Researchers explored Inchon International Airport by using a virtual airport model that measures the quality of airport service as well as other factors that have an influence on transfer passengers' behavior.

An airport facility will effectively ensure that the airport offerings are positioned uniquely so that the airport staff can maintain a high level of service quality. The electric taxi system can take the aircraft to areas of the airport that were previously unobtainable using main engine thrust taxi. Passengers will notice the flexibility that the electric taxi system adds when making the best use of their time spent at the airport.

Flight safety could be influenced by a combination of factors, including pilot operations, airplane environment conditions and organizational situations. Most commercial flights can be generally divided into five parts, boarding, takeoff, time at cruise, landing, and deplaning. Statistics will show that most aviation accidents happen during the takeoff and landing. Accident investigations show human errors are usually the main cause for these accidents. The aircraft electric taxi system will have a direct effect on four of these five parts: aircraft taxiing to the gate before boarding passengers, taxiing away from the gate into position before takeoff, and taxiing to the gate after landing for deplaning. It will be interesting to see how the ground safety improves after the first few years that the aircraft electric taxi is in service on commercial airliners.

In the commercial aviation industry, the black box or flight data recorder is used as the front-line recording equipment for aircraft operational data and is used widely to investigate incidents or accidents. The Civil Aviation Administration (CAA) of China requires that all aircraft operated commercially in China have a quick access recorder installed. The quick access recorder can record hundreds of aircraft operational parameters at the same times. This tool can provide a great number of recorded flight data and makes reliable risk analysis possible. This data may also be useful for an airline when it is developing a business case for implementation of electric taxi system.

Aviation safety has gone through three stages: the human period, the machine period, and the organizational period. Currently, aviation safety management practices are shifting from a reactive mode to a proactive mode. The modern commercial aviation industry is busy with on-condition maintenance, which involves fixing or replacing components before there is a problem. The aircraft electric taxi system offers a large increase in safety factors due to the main engines being shut down during taxi operation. The installation of the aircraft electric taxi systems on commercial aircraft could be seen as being proactive for the safety of the ground support personnel.

There are a large number of ground operations and flight data that exceeds the normal limits when the flight performance data recorded by the quick access recorder is analyzed. Most of the over limit operations, such as hard landings or taxiing the aircraft into places it doesn't belong, are caused by the pilot. Operating aircraft with over limit conditions could lead to very serious consequences especially during inclement weather.

Some industry experts believe that there is an increasing need for improved risk analysis methods to evaluate the safety and risk of pilot decisions. They provide a new method to evaluate pilot operations and give solid proposals for improvement. Furthermore, the quick access recorder data has been used to diagnose and improve ground and flight operations.

The quick access recorder information is used as part of the flight operations quality assurance program among the general aviation industry and has been well-received. Some industry experts have established neutral networks with the quick access recorder data to analyze the severity and causes of ground operations and flight over limit operations, such as touchdown point and hard landing failure. Hard landings can be an over limit condition, as it occurs most frequently. Some hazardous situations during pilots' operations for this over limit condition can suggest instructional avenues for future pilot training.

The air traffic control system is a set of ground-based services provided by air traffic controllers. The air traffic controllers communicate with the aircraft in order to prevent mid-air and on-ground collisions by separating aircraft and supplying relevant information and advisories for safe aircraft operation both in the air and on the ground. Air traffic control centers and airport ground control are saturated, and their capacity would be exceeded if flights were not regulated in advance, especially since airline ground traffic and airline air traffic at most major airports has increased for several decades. With the adoption of the electric taxi system, ground controllers will be able to direct aircraft ground traffic using their electric taxi system to places on the airport that were not previously accessible using main engine thrust for taxiing.

For more than 40 years, global air traffic has also continually increased. However, this growth has not been followed by an equivalent increase in the capacity of air traffic control and ground traffic control, which leads to a severe increase in congestion both at airports in the United States and Europe. The saturation of the air traffic system that includes aircraft control on the ground greatly affects these airports due to the distinct nature of air traffic. One way of increasing capacity is to use present resources more effectively. The adoption of the aircraft electric taxi system would give better access to use these present airport resources.

Air traffic management in Europe is structured by several layers of filters working at different time horizons. The strategic filter is concerned with long-term air traffic management and collects all activities related to the management of airspace, such as the definition of the airspace volumes, the geometry of air routes, and the climb and descent procedures around airports. The pre-tactical filter is activated a few days to a few hours before the scheduled traffic implementing regulation measures, in order to adapt the traffic load to the airspace capacity defined during the strategic phase.

Next, the tactical filter corresponds with air traffic controller traffic management to monitor the traffic, coordinate the traffic between control sectors, and resolve conflicts. Last, the emergency filter is triggered in case of a failure of all the preceding filters. The ground-based and airborne systems warn controllers or provide maneuvers to pilots to prevent an accident if a collision is imminent between two aircraft or between an aircraft and a ground obstacle. Aircraft equipped with the electric taxi system will have improved mobility and the ability to maneuver through the airport ground obstacles with a lower of risk of unintentional contact, which will also lower lengthy tarmac delays created by unintentional contact.

In 2009, the U.S. Department of Transportation (DOT) issued a final ruling that mandates U.S. carriers adopt alternate plans for longer than usual tarmac delays. These alternate plans include a guarantee that an airline will not allow their commercial aircraft to stay out on the airport tarmac longer than two hours for domestic flights in the U.S. when passengers are on board. For international flights, the rule does not apply unless it is in the selling carrier's U.S. Contract of Carriage. In the case of international travel, the length of time on-board before deplaning is determined by the U.S. Carrier (14CFR 259.4). There might be exceptions to the rule for safety, security, or air traffic control-related reasons.

This ruling, which went into effect in 2010, is referred to as the tarmac delay rule or phase one of the U.S. Department of Transportation consumer protection efforts. This phase one is now the law and air carriers in the United States have to abide by it. Around eight weeks after phase one went into effect, the U.S. Department of Transportation started phase two. Phase two uses a notice of proposed rulemaking. In the aviation industry the notice of proposed rulemaking is only a white paper published for the public to review and give feedback to the Department of Transportation if needed. It is not a rule at the time of publishing.

If phase two is put into effect, the rule would apply to foreign airlines as well. Phase two could force more airlines to be compliant and substantiate additional tarmac delay requirements on a wide range of other issues. These issues could include denied boarding reimbursement, full-fare airfare advertising, and allergies related to peanuts. The U.S. and foreign airlines are against the phase two rule, which remains pending.

Airline management should focus on effectively managing their operations to provide good quality and maintain a high level of aviation safety. It is unfortunate that flight schedules and flight time departure accuracies are seen as unpredictable. The flight schedule sometimes becomes unobtainable because of disruptions. These disruptions can be caused by bad weather conditions, unpredicted maintenance requirements, and earthquakes, among others. The aircraft electric taxi system can be seen as a high-quality service addition and an increase in ground safety when aircraft are equipped with it. For while it cannot eliminate disruptions, it can help to mitigate the risk of their effect.

Airline schedules can be eroded by large-area flight delays that occur when the magnitude of the flight delay exceeds a certain level. Customers' trips are delayed and sometimes cancelled when large area flight delays occur, making large area flight delays a key factor when trying to maintain customer satisfaction.

Airline companies are faced with the threat of huge financial losses and a reduction in the quality of services when flight delays occur. Yet, airlines face a venture risk that could be addressed with cost effective management methods. Flight delays have become the major obstacle impeding the expansion of the global aviation industry, especially the expansion of the worldwide aviation community into emerging countries and the increasing variety in aircraft models comprising airline fleets.

Airports are the aircraft control system bottleneck where most flight delays occur. Aircraft are currently pushed out from the gate with a ground support tug for the first few hundred feet. The pilot starts the main engines and taxis away using main engine thrust, along with all the hazards and safety concerns related to ground support personnel, gas turbine engine foreign object damage, and objects on the ground affected by the engine exhaust blast. One of the features and benefits of the aircraft electric taxi system is it allows the aircraft to taxi away from the gate using electric power. It does not have to be a pilot who operates the aircraft under electric power; a trained ground crew member can do it. The ground crew member can taxi the electric taxi equipped aircraft off to a non-gate related place for between flight servicing, freeing up the gate area, then taxi it back to the gate when completed.

Aircraft recovery, crew recovery, and integrated flight rescheduling in the airline industry is primarily about disruption management. A great deal of research focuses on flight postponement management methods. For instance, one study provided a general preamble to airline disruption management and includes a description of the planning processes used in the airline industry, including optimization and replication models for airline customers and flight crew recovery.

The method of quantifying airport accessibility can scientifically reflect the degree of convenience for airport passengers. Research has shown the higher the level of accessibility, the stronger the competitiveness of the airport will be. The accessibility of the airport has a significant relevance with indexes such as passenger scale and number of airlines. When the level of airport accessibility rises by 1%, the passenger volume will increase by 2%.

There are several threats connected with the introduction of electric taxi systems to aircraft operations, and future electric taxi system studies are recommended to predict the developments and potential impacts on air operators and airports. However, as this section shows, there are positive economic impacts of using an electric taxiing system.

The Cost of Aircraft Engine Maintenance Due to Foreign Object Damage

Damage caused by foreign object debris at airports can cost the airport tenants and airlines millions of dollars every year. When gas turbine engine equipped aircraft taxi using main engine thrust, the air intake at the front of the engines acts like a giant vacuum cleaner sucking in air as well as anything on the ground that happens to be in the low-pressure area caused by the turbine intake. Items that are in the range of this negative pressure area are called foreign object debris. According to The Boeing Aircraft Company, foreign object debris can be anything that does not have a purpose in or near an airplane and, as a result, cause damage to airplanes and injure airline or airport personnel. Foreign object debris damage can be caused by loose hardware, building materials, sand, rocks, dead animals, and pieces of luggage. Foreign object debris is found at taxiways, cargo aprons, terminal gates, aircraft run-up areas, and runways. Damage caused by foreign object debris is estimated to cost the airline industry over $4 billion a year. Airport tenants and airlines can take steps to reduce this cost and prevent this type of damage.

Foreign object damage can also result from objects being caught up in the jet blast and thrown into unpredictable places. The effect of the unplanned damage from foreign object debris can have a catastrophic impact on maintenance costs. For example, to repair an engine damaged by a foreign object could be greater than $1 million. Foreign object debris damage can be a major cause of increased indirect costs through flight delays, cancellations, and loss of revenue. There is also potential liability due to additional work and injury for airline staff and management.

Table 2
*Purchase Price and Cost to Repair FOD Damaged Aircraft Engines_*

Engine Type Costs

Purchase Cost of MD11 Engine $8 – 10 Million

Purchase Cost of MD80 Engine $4 – 6 Million

MD-11 engine overhaul to correct FOD damage $500, 00 - 1.6 Million

MD-80 engine overhaul to correct FOD damage $250,000-1.0 million

MD-11 fan blades (per set*) $25,000

MD-80 fan blades (per set*) $7,000

*Fan blades are balanced and replaced as a set.

The examples shown in Table 2 indicate the cost incurred by the airline to repair a foreign object debris damaged engine can exceed 25% of its original purchase price. The amount of money spent to repair foreign object debris damage each year is very high. Most of this debris is ingested by the aircraft during taxi. In most cases, the actual time and place of the foreign object debris damage cannot be pinpointed. Much of the damage caused by foreign object debris is not discovered until the engine goes in for heavy maintenance at regularly scheduled intervals spelled out by the operator's maintenance manual. When the individual aircraft has a record of airport visits, taxiways traveled, and time on the ground with engines running, the location of the actual foreign object debris damage could be easier to identify.

The risk of foreign object debris damage is mitigated by the electric taxi system. Aircraft that are equipped with the electric taxi system can taxi or move about the airport tarmac without the main engines running. While the aircraft is taxiing using electric power, it will reduce if not eliminate the chances of foreign object debris being ingested by the main engines during ground operation.

The Future of the ETS Equipped Single Aisle Narrow-body Aircraft

There is a strong future for the single aisle narrow-body aircraft (SANBA), but there are many measures that can affect the decision to promote the design of a new aircraft. Industry experts focus on whether or not to take on the high costs of development in changing the structure of the airplane model. Once the aircraft manufacturer has certified a new model of aircraft for scheduled commercial airline service, it can take them up to and maybe more than twenty years to recover their initial investment.

A good example would be the Boeing narrow-body 737. This model aircraft went into service in the 1970s. This aircraft is still being produced in 2020 with minor modifications, not changing the structure of the original model. The initial target market for the aircraft electric taxi system manufacturers is the Boeing narrow-body B-737 and the Airbus family of A-320 narrow-body aircraft.

It is possible that there will be two main problems regarding the aircraft electric taxi system certification for use on commercial aircraft. The first is connected with methods of fuel planning. Rules may prohibit fuel planning based on the operation of the electric taxi system. It may be required to carry taxi fuel for worst-case scenarios when the main engine thrust taxi systems are needed. There may also be changes in the minimal aircraft operational equipment list after the implementation of the electric taxi system.

Another issue is the use of the electric taxi system under poor airport tarmac friction situations. It is important for every airport to choose their friction limits for the times when the electric taxi system will be prohibited since it would be very difficult to control ground movements of rapidly accelerating aircraft because of a high skid. It may be a suitable idea to prohibit the use of the electric taxi system while the friction coefficient remains low. Moreover, there will probably be some airports with prohibited areas, such as taxiways with a great slope or taxiways where it is necessary to cross an active runway.

The electric taxi system will also lead to necessary changes in airport design. It may become common to place a stand still area near a holding point for aircraft that experience problems during their engine start up procedure. On the other hand, changing the layout of standstill areas may also be beneficial because electricity-propelled aircraft will need less space to operate during pushback. With regard to airport handling operations, it is important to note that there may be savings when it comes to investments towards towing tractors since it is possible that their numbers will be reduced.

The electric taxi system has been originally designed for aircraft the size of the A-320 or B-737. Development for smaller aircraft types depends on whether they have large batteries or auxiliary power units installed. Smaller aircraft without auxiliary power units installed often use one engine as an electric generator with a braked propeller. Such systems generate electric energy less efficiently, and so the use of the electric taxi system is not as effective and not as attractive without a larger and much heavier battery. Larger aircraft with a greater taxi mass might not have enough auxiliary power to spare for the electric taxi system, especially during hot days when their air conditioning systems are operating at maximum capacity. These airplanes will have to be assessed for electric taxi system compatibility. Another possibility is to use the electric taxi system for helicopter operations.

The savings when the electric taxi system is used may be so significant that it could result in new aircraft designs. It is possible that after a few years of successful electric taxi operations, there will be new versions of current aircraft types equipped with an electric taxi system as an integral part of the actual aircraft production. Furthermore, there may be new aircraft versions equipped with more powerful auxiliary power units, or larger batteries, designed exclusively for electric taxiing.

Some uncertainties remain which will make airlines cautious. One of these is the fact that there is no proven, official information about the electric taxi system performance. This includes the performance of these systems at airports with a difficult incline layout. Furthermore, there will be a decrease in the system performance for lower friction, so it is possible that operational manuals for some airports and some weather situations will prohibit the use of electric taxi systems. Some airports may limit the use of the electric systems when leaving or crossing runways for safety reasons. Air operators will probably wait for the first operational statistics before deciding to obtain these systems.

Airplanes have progressed technologically from the banks of radial piston engines to the flat piston engine to the new jet engine adaptability design. In the early 1960s new production methods enabled new aircraft designs for greater speed and capacity. There was rapid airplane technology development in the 1980s as well. This was due to new propulsion systems, the introduction of new materials, and new technology developments with electronic instruments. Fuel savings increased as did reliability, speed, and safety. Through these historical developments, moving the aircraft around the airport tarmac was totally reliant on main engine thrust or tractors to tow the aircraft when needed.

Further new manufacturing methods were implemented in the 1990s after being developed in the 1980s. During the 1990s, industry experts thought that airplane technology had matured. Yet, more recently aircraft technology has been transforming rapidly. These changes focus on information technology, increased aerodynamics, design tools, increased range, electronic control systems, and steel structures replaced by the rapid growth in composite material technology.

The next technological breakthrough in air travel as seen by the public, which will be part of the aviation historical development of disruptive technology, will be aircraft taxiing on the tarmac without the use of the aircraft' main engines. The application of the electric taxi system will facilitate development, beginning with the narrow-body aircraft. Industry experts say that the single aisle narrow-body aircraft industry has been the most profitable for airlines and aircraft engine manufacturers. The future for this platform of aircraft looks good, and the electric taxi system will only contribute to its continued positive performance and profit.

Landing Gear Compatibility for the ETS Installation

The landing gear currently manufactured for the narrow-body aircraft are strong enough to support the new electric taxi system kit as offered from the systems manufacturers. The engineers who are currently sponsoring discussions of the Air Force Heavy-Press Program for light metals will need to analyze and approve the electric taxi systems if the aircraft manufacturers decide to modify the landing gear from its original design to better suit the electric taxi system component installations while the aircraft is on the production assembly line. The program involves the design and building of huge presses used for forming very large integral sections of airframes such as wing and fuselage members, as well as landing-gear components and other parts. Such structures are essential for the high-strength and lightweight requirements of present and future supersonic aircraft. Forgings and extrusions are most suitable for making these parts, which must have precision of form, great strength, be light weight, as well as be constructed for the lowest possible cost.

Drawbacks and Complications Related to ETS Installation and Operation

All commercial aircraft are required to undergo a 100-hr. inspection as prescribed by the manufacturer and approved by the Federal Aviation Administration. Aircraft equipped with the electric taxi system will require additional down time (removed from revenue service) to service the landing gear and related pilot controls as recommended by the manufacturer every 100 hrs. These inspections and service requirements have not yet been identified. The electric taxi system kit manufacturers will have to specify if inspections can be performed at the gate during an overnight stay at the airport or if the aircraft must return to a fully equipped maintenance hangar for attention.

Logistics support will need to keep enough spare parts available at all airports where electric taxi system equipped aircraft are used. The airline parts aftermarket suppliers will have to keep an inventory ready to airfreight to the airlines to support electronic taxi system operations. The cost of spares support could outweigh some of the advantages of fuel savings.

For inspection, the electric taxi system could be part of the airline minimum equipment list (MEL) or master minimum equipment list (MMEL). The MEL is a list of aircraft components that are required to be operational in order for the aircraft to operate safely. An example of this could be some of the older B-737 aircraft, which have collapsible stairs that fold up into the passenger doors. If the stairs are not operational and the airport does not have portable stairs, the aircraft cannot be used for revenue flight at the particular airport.

The same case scenario could go for an aircraft equipped with the electric taxi system that is not working. Even though the aircraft can still taxi using main engine thrust, the airport facilities might have been modified so main engine thrust taxiing is not allowed, due to noise and safety restrictions. In this case only electric taxi system equipped aircraft could access the terminals.

A potential drawback of the electric taxi system is the added weight of the system. The electric taxi system weight varies depending on manufacturer, but estimates are between 150 and 450 lbs. This additional mass has to be transported all the time, which leads to additional fuel consumption. Less fuel used for taxiing and carried by the aircraft compensates for this additional weight. However, it is likely that the aviation authorities will impose strict limitations on the operation of aircraft regarding the fuel quantities required by a working electric taxi system. Another reason to carry additional fuel is if the aircraft is diverted to an airport where electric taxiing is prohibited.

Another issue to consider is the potential traction problem with aircraft tires due to the electric taxi system installation. As the traction is transferred to the drive train, there is potentially greater wear on the tires. Both the hysteresis wear and abrasive tear due to skidding have to be taken into account for the power driving wheel. For non-driven wheels, there may be additional wear caused by more acute maneuvers allowed by the electric taxiing system, which will cause more tire dragging. Maintenance and replacement of those tires may then have to change. The cost of tires depends on the aircraft type, the tire producer, if the tire is mounted on the main or front wheel, and if the tire is retreaded or brand new. In the worst case, the main wheel tires on a

Boeing 737 cost up to $13,000 each. The shortened life of the aircraft tires will have to be accounted for in the electric taxi system cost of operation.

Improvement of Airport Area Air Quality

Aircraft equipped with the electric taxi system will contribute less to local airport area air pollution as the aircraft carbon footprint is reduced. This contribution should not be underestimated as continued public awareness and growth in air traffic has made the environment a main focus in the future of the commercial aviation industry. It is widely known that sustaining the long-term growth of air transport depends upon environmental improvements for future acceptability.

During aircraft taxi, the release of exhaust gasses in the atmosphere on the airport tarmac is the third most important environmental issue related to commercial airliners. These gasses can be seen at most major airports as a brown haze hanging about 500 to 1500 feet above the airport and surrounding areas. The anticipated doubling of airline fleets in the next twenty years will certainly bring the issue to the forefront. More fuel-efficient aircraft engines that emit fewer pollutants are currently in design by most jet engine manufacturers and will help offset the increase from growth in air traffic. The contribution of aviation emissions is expected to be increased by a factor of 1.6 to 10, depending on the fuel use scenario.

In theory, aircraft emissions have declined over time when considering the emissions from transporting one passenger one mile. The overall increase is from a system wide increase in aircraft passenger capacity. Current air quality regulations have focused on local emissions generated in airport vicinities. Aircraft operating cycles are usually identified by two major parts. The Landing-Take-off (LTO) cycle which includes anything near the airport that takes place below the altitude of 3000 feet (914 m) above ground level, this includes taxi-out and in, take-off, climb-out, and approach-landing. With the application of the electric taxi system, air pollution created while taxiing can be eliminated and fuel consumption reduced.

Reducing CO_2 Emissions

The ETS Equipped Aircraft Cuts Emissions in the Takeoff Queue	
Per flight fuel savings for a 5-minute queue (assumes dual-engine @ 2.9 gal/minute)	14.5 gal
Daily fuel savings, one plane (4.5 turns)	65.3 gal
Daily fleet savings (100 planes)	6,500 gal
Annual fleet savings (360 days)	2,350,000 gal
Annual reduction in CO_2 emissions @ 9.6 kg CO_2/gal Jet A	~ 22.6 million kg

(Gallons shown are US. Savings in de-icing queues not included.)

The amount of taxi time varies according to airports. One example, however, comes from a study of commercial jets at Boston Logan airport that spent 30% of the total time taxiing on the tarmac before takeoff. This does not include the taxi time to the terminal after landing. Taxi out time averages around 19 minutes at Boston Logan airport, with an average of 247 flights per day. In the United States, aircraft taxi time translates into 6 million metric tons of carbon dioxide, 45,000 metric tons of carbon monoxide, 8,000 metric tons of nitrogen oxides, and 4,000 metric tons of hydrocarbons emitted into the atmosphere on an annual basis.

With the application of the electric taxi system at Boston Logan airport, about 36 total hours of running the main engines can be eliminated, which results in a savings of 7,800 to 9,800 gallons per year. With the average cost of jet fuel estimated at $3.50 per gallon, this equates to a savings of over $30,000 per day, if all 247 aircraft have the electric taxi system installation.

The electric taxi system can be considered green technology as it uses batteries or an auxiliary power unit for electrical power instead of the main engine for propulsion during ground taxi operations. The auxiliary power unit must be operated to provide cabin air conditioning and functions more efficiently than main engines while in ground mode. This optimal operating together with the environmental, social, and economic sustainability achieved with an electric taxi system makes a positive contribution to the communities that airlines and airports serve.

Since transport systems exist to provide opportunities for people to make social and economic connections and since the public increasingly values environmentally friendly practices, consumers may see electric taxi system installations as a step towards saving the environment.

Employees and consumers look for companies that are environmentally friendly in the way they operate and the products they make. Furthermore, businesses are discovering that day-to-day cost savings can be realized when steps are taken towards saving the environment. In addition, these efforts improve company employee recruitment and retention and help the local environment. Airlines that install the electric taxi system to enable green taxiing are setting a good example for their employees as well as their customers.

Real time pollution reduction evaluations can be conducted and advantages determined once the first few electric taxi system equipped aircraft are in service. As this chapter has noted, aircraft currently move across the airport tarmac using the thrust of the main engines or by tow with an airport tug, burning much more fuel than is necessary. Using the aircraft's electrical system, sustained by the auxiliary power unit to drive electric motors, is preferable.

This configuration would keep the movement of the airplane independent of a tow vehicle, consuming less fuel and producing fewer emissions. One fact is interesting: when the aircraft has an auxiliary power unit to charge the batteries and power the electric wheel drive motors, the gas turbine powered auxiliary power can be shut down during ground taxi and rely totally on the aircraft batteries to save fuel and to eliminate all pollution generated by the aircraft.

The United Kingdom has set a target for overall CO_2 reductions of 80% by 2050 relative to the 1990's CO_2 measurement levels. Aviation contributes to only about 6.3% of the United Kingdom's carbon emissions. While it may be argued that this impact is low, aviation in the United Kingdom is growing at approximately 8% annually, and its projected that growth will make aviation carbon emissions a primary concern.

Most of the United Kingdom's allowable CO_2 emissions will be derived from aviation by 2050. This suggests that aviation will remain in the spotlight for the foreseeable future, and against this background, every sector of the industry will have to act to minimize CO_2 emissions. The adoption of the electric taxi system by United Kingdom based airlines could help reduce ground-based emissions by 4% annually.

The benefits will not be for the United Kingdom alone. Another study presented a detailed estimation of fuel consumption and emissions during taxi operations using aircraft position data from actual operations at Dallas/Fort Worth International Airport. The study assessed thrust level during each state, fuel flow, and emission index values from International Civil Aviation Organization's databank. Ultimately, the study provided a relative comparison of all the taxi phases and their contribution to the total fuel consumption and emissions during main engine thrust taxi. Stop-and-go situations, resulting primarily from congestion on airport's taxiway system, account for approximately 18% of fuel consumed. Aircraft gas turbine engines are designed and certified for optimum fuel consumption and predicted emissions performance at cruise speed. During the engine certification process, the emissions at idle data are recorded, but there is no minimum or maximum emissions requirements. The states of idling and taxiing at constant speed or braking were found to be the two largest sources of inefficient fuel burn and emissions.

Potential Costs of Electronic Taxi System Installations

There are currently two aircraft parts manufacturers who are qualifying their electric taxi system kits for Federal Aviation Administration regulations: Safran Aerospace and WheelTug plc. While there has been a request for information, along with letters of intent from over 50 major airlines, neither of these companies will quote a price until an airline asks for a quote. To date, no one has requested a quote.

Airline industry experts forecast that the electric taxi system will be ready for production and aircraft retrofit installation in 2022. The predicted cost to retrofit the most common narrow-body Airbus A-320 or Boeing B-737 will have to show a positive return on investment (ROI) in less than three to five years to be an attractive investment for airlines. The cost to install and maintain the electronic taxi system will be amortized over 20 years, which is the revenue life of the aircraft.

A cost benefit analysis highlights the estimated positive and negative impacts of electric taxiing systems. This analysis shows that savings may be achieved during every flight, including the most important savings of fuel and maintenance. The greatest risk is delay because of the different taxi out times of electric taxi system equipped and unequipped aircraft.

The aircraft electric taxi system is a promising new technology for the airline industry. Currently there is little literature about it, and more information is needed to allow the airline industry to make a decision on how to best implement the installations of the electric taxi system.

Pushback Benefit vs. Today

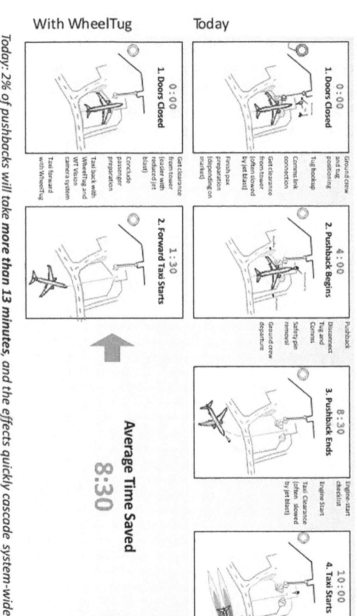

Today

0:00 — 1. Doors Closed
- Ground crew and tug positioning
- Tug hookup
- Comms link connection
- Get clearance from tower (often slowed by jet blast)
- Finish pax preparation (depending on market)

4:00 — 2. Pushback Begins
- Pushback
- Disconnect Tug and Comms
- Safety pin removal
- Ground crew departure

8:30 — 3. Pushback Ends
- Engine-start checklist
- Engine Start
- Taxi Clearance (often slowed by jet blast)

10:00 — 4. Taxi Starts

With WheelTug

0:00 — 1. Doors Closed
- Conclude passenger preparation
- Taxi back with WheelTug and WT Vision camera system
- Taxi forward with WheelTug
- Get clearance from tower (easier with reduced jet blast)

1:30 — 2. Forward Taxi Starts

Average Time Saved 8:30

Today: 2% of pushbacks will take more than 13 minutes, and the effects quickly cascade system-wide.

CHAPTER 3:

Airline Operations

The aircraft electric taxi system kit manufacturers suggest that an aircraft using the electric taxi system will be able to push back from the terminal gate in less time than an aircraft using ground support tugs or reverse thrust from the main engines.

The aircraft electric taxi system developed by WheelTug plc uses electric motors in the nose wheel. The manufacturer claims that pilots can back planes away from gates without a tow tug and then taxi to a remote location on the airfield before starting the main engines. After landing, the pilot can turn off the main engines and switch to electrical power from the aircraft battery and auxiliary power unit to operate the motors. Increasing efficiency of movement of the aircraft from the gate to the airfield can reduce the tarmac congestion as well as reduce fuel consumption.

There are reasons that airlines might be unwilling to adopt the electric taxi system including 1) that the aircraft electric taxi system is seen as a major revolution in aircraft ground handling operations; 2) additional training will be required to operate the aircraft under electrical power; 3) the increased ground maneuvering efficiency due to the electric taxi system installations could make the non-electric taxi system equipped aircraft seem disabled or obviously inefficient; or 4) there is a hypothetical risk that the electric taxi system can create logistical problems resulting in flight delays, cancelling out any value added benefits.

The necessity for an optional system for aircraft movement that decreases fuel burn, lessens engine upkeep costs, and conveys environmental benefits is now more noteworthy than at any other time in recent history. The answer for this issue might be the use of alternate means to taxi aircraft. The electric taxi uses electrical power from the aircraft's battery or auxiliary power units to power motors connected to the landing gear apparatus wheels. Consequently, the electric taxi system application permits an aircraft to leave the terminal gate without any ground-based aid, then taxi to the runway before starting the aircraft main engines.

Industry experts say the electric taxi system can be installed on the aircraft in two overnight maintenance visits (150 man-hours), as it is retro-fit able and removable. The electric taxi system is multi engine minimal equipment list exempt. This means the aircraft can still perform commercial service when the electric taxi system is not operational. With the ample power from the auxiliary power unit and/or aircraft batteries (not affected by engine start), the pilot moves forward and reverse using the control panel and standard tiller to steer. Initial pilot training would be 45-60 minutes.

Potential Aircraft Installation Population

There are approximately 400 airlines operating globally. In the United States, there are 173 airlines. The core focus for the electric taxi system is American Airlines and United Airlines, which both operate Airbus and Boeing narrow-body aircraft. These models of aircraft see the most cycles, which is defined as one take off to altitude cruise using full cabin pressurization followed by descent and landing to a full stop, on an annual basis and are the target for the operational cost savings.

Operational Costs

The cost per operating hour and fuel savings per hour was calculated using information from the aircraft electric taxi system kit manufacturers. The aircraft manufacturer's certification data such as fuel consumption during taxi and the day price of jet fuel was used to establish a baseline for the cost to taxi an aircraft using main engine thrust. The engine manufacturers' certification data that includes all engine performance information was analyzed. Review of the data substantiated the average time the aircraft spends at idle (7% taxi power) and the amount of fuel consumed. The cost of the electric taxi system installation were amortized over the revenue life of the aircraft which is 20 years, see Appendix A for additional airline specific information.

Some airlines might require a quicker payback for the investment. This ROI and payback analysis is broken down for discussion in Chapter 4. The number of years to recover the electric taxi system cost through fuel savings alone is the primary focus for this discussion. Increased airport facility use, reduced labor requirements, and increased flight schedule flexibility are discussed but the actual cost savings are not detailed at this time. The dollar amount in savings from these added benefits can vary from airline to airline depending on the particular business model.

For some airlines, the cost of fuel might not be a measurable factor as some are part of an alliance where fuel purchases are forecasted and contractual. In these cases, other features and benefits of the electric taxi system installation will have greater importance. These measurements can be calculated with the benefit of all aircraft operational savings, less the price of jet fuel.

As part of the aircraft electric taxi system installation and aircraft operational cost reduction, we identified all airplane traffic movements on a typical day at Phoenix Sky Harbor airport. This information was available at the airport management office located next to Terminal 2 at Sky Harbor Airport. The operational costs were calculated by the number of engines per aircraft and the numbers of aircraft, then multiplied by the typical time the aircraft engines are operated during taxi and ground operations.

Aircraft operate in taxi mode for 26 minutes between landing and gate arrival, and from gate departure to takeoff, as estimated in the engine manufacturer's certification data. The current price of jet fuel and other cost saving features after the electric taxi system installation were used for the cost savings calculation (see Appendix A).

The scope was limited to the study of cost savings related to the ground operations of the Boeing 737 and Airbus A-320 family. The fuel consumed while the aircraft is taxiing on the ground was measured and quantifiable. The material costs to purchase the electric taxi system were identified by the industry for this narrow-body aircraft. The lager twin aisle aircraft were left out of this study because current aerospace technology has not addressed an applicable electric taxi system.

Airport area air quality improvements resulting from alternate forms of taxi were mentioned but not analyzed.

Two major U.S. based domestic airlines, United Airlines and American Airlines were chosen for this study's focus. These airlines operate the same make and models of narrow-body aircraft and perform the same in-house aircraft fleet maintenance. Their cost to operate and maintain their aircraft should be about the same.

Airlines operate with high fixed fuel costs and a limited ability to respond to irregular or changing travel patterns. In addition to fuel, whether aircraft are owned or leased together with operating and maintaining costs are significant expenses for airlines. The goal of most airlines is to have planes operating with maximum fuel efficiency and full of passengers as many hours of the day as is practical.

In today's era of exorbitant airport charges and rising fuel prices, it is important for airline companies to develop ways to cut fuel costs. In order to cut fuel costs and increase performance, airlines work constantly on reducing turnaround times. These smaller and larger adaptations can be seen as important process innovations, and they are crucial for competitiveness.

The limited nature of oil, and hence aviation fuel, is increasingly becoming a restraining factor for profitability in the air transport industry. Aircraft technology and design, aviation operations infrastructure, socioeconomic and political measures, alternative fuels and fuel properties, and aviation infrastructure are the five key dimensions affecting the cost of fuel and efficiency.

The airline industry faces substantial financial risk exposure that affects the vulnerability of stock returns which arises from changing economic conditions, volatile fuel price movements, and fluctuations in exchange rates. These exposures are attributed to the cyclical demand, strong price competition, high gearing levels, capital investment, fixed costs of labor and equipment, and regulatory impediments such as ownership restrictions and landing rights.

Working capital, like cash and liquid assets, runs the facilities and supports the daily activities of airlines. This kind of capital is essential for continuing activities and increasing volume. It also ensures airlines maintain credibility, reduce risks, and can overcome extraordinary situations. Working capital management and return on investments are important for making profits, especially for dynamic sectors like aviation. An airline must regularly re-evaluate its business model against the options available and invest time, effort, and money to move from one mode of revenue management to another.

Developing new business lines is assessed by estimating the difference between the existing and new lines, including their impact on organizational performance and additional indicators such as financial effort, the estimated return on investment, technical and organizational difficulty, risk level, and domain financing opportunities. Every entrepreneur, when setting up or developing a business, envisages its sustainability, its ability to continue for a long time. Sustainability implies return on investment in three major aspects: environmental, social, and economic which have to be carefully balanced. For airlines, innovation is strongly related to sustainability and should focus not only on products and services offered, but on all of their key business processes.

The airline industry is one of the most important sectors worldwide because of its global nature. In recent years, the airline industry has experienced new conditions of liberalization: increasing competition, economic and traffic growth, acquisitions and mergers, bankruptcy, volatility in earnings, considerable profits and losses, innovations, and the emergence of low-cost carriers. Due to global competition in the new era, corporate finance and return on investments play an essential role in maintaining efficient airline operations in short and long-term decision-making and results. Return on assets as well as return on investments highlight how efficiently assets are used to generate earnings.

In the United States, there have been government sponsored tax incentives for using biofuels under the Environmental Protection Agency's Renewable Fuel Standard program. These are subject to change with each federal budget cycle. There are also federal subsidies and loans given to renewable aviation fuel producers. However, the availability of cheap oil, currently around $50 per barrel, reduces the incentive for commercial airlines to use biofuels beyond the level incentivized by tax relief.

In the United States, airlines are private enterprises. They use their own capital to run operations. However, these carriers also receive economic relief from states. The U.S. airlines have collectively secured tax breaks worth billions of dollars from state and local governments. The aircraft engine performance data in table 3 shows the amount of jet fuel used during aircraft main engine thrust taxiing operations.

Table 3

Aircraft Engine Performance During Taxi Operations

Fuel Burn CFM-56 engine 850 lb. hr. 14 lb. min. X 54 min. = 756 lbs.

6.5 lbs. per gallon / 756 lbs. = 116 gal. X 2 engines = 232 gal.

Cost of fuel consumed during taxi. $329.44 ($1.42 X 232= $329.44)

http://aviation.stackexchange.com/questions/ 8429/does-the-apu-consume-a-lot-of-fuel-compared- to-a-jet-

---

CHAPTER 4:

Conclusion

The total estimated cost to support one aircraft cycle, from the time the aircraft leaves the runway after landing, then taxis to the passenger gate, then taxis back to the runway threshold ready for takeoff is $329.44. The total number of United Airlines A-320 fleet cycles is 197,936 a year. The number of cycles calculated per aircraft (155 /197,936) is 1,277 per year.

The ETS Equipped Aircraft Cuts Cycle Times
And increases flights per week

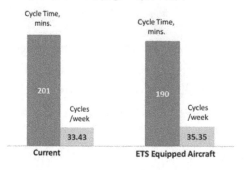

When examining the sample case of United Airlines who currently operates a fleet of 155 A-320 aircraft, there is a potential annual fuel savings of $65,207,706 if the aircraft electric taxi system technology is applied to the entire fleet. In this case, fuel consumption and ground support personnel required at the terminal gate would be reduced. The equipment needed at the gate would also be reduced as there is no need for the aircraft tug when using the electric taxi technology.

The ETS Aircraft Increases Fleet Size

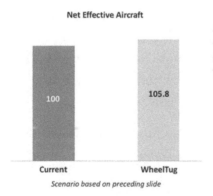

Net Effective Aircraft

100   105.8

Current   WheelTug

Scenario based on preceding slide

An airline only has to decide how many destinations to add to the schedule!

For this discussion the latest published cost for the electric taxi system kit is $925,000. For these calculations, however, we will round up the cost of the electric taxi system up to $1,000,000 and use the aircraft ground performance data from Table 3. This additional 8% could take into account the airline's cost, labor to install, taxes, or other incidentals. That means that the annual savings per aircraft after the electric taxi system technology is adopted would be $420,694.88. When airlines invest $1,000,000 to adopt the electric taxi system technology with one aircraft, they should see a payback ROI in 2.4 years ($420,694.88 / $1,000,000 = 2.4).

The electric taxi system manufacturer WheelTug claims that the time saving will be a more important issue for electric taxi users.

The ETS Equipped Aircraft Enables More Flights

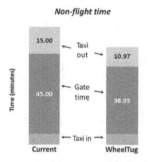

Non-flight time

The ETS equipped aircraft pushback times are not only shorter, they are also more dependable, requiring less schedule padding by airlines to meet on-time metrics.

With these efficiencies, the ETS equipped aircraft can open up significant schedule gains for any airline.

Savings in this scenario (year 5) are 11.01 minutes/flight of *scheduled* time. Savings from two-door (Twist) handling included.

*Depending on an airline's current fleet and schedule, WheelTug productivity gains may increase **10%** or more.*

They propose that summing minute-savings from several flights can create a time slot for another flight. This assumption requires the flexibility of airline operations, allowing the use of these time savings.

Some airline professionals present a detailed analysis of the positive and negative aspects related to the electric taxi system. They focus on the economic impacts of using an electric taxi system and evaluate the possible savings. While there may be potential threats connected to the introduction of these systems, future electric taxi systems modeling could predict the developments and potential impacts on air operators and airports.

Aircraft Utilization Benefits

- In this model, available seat km increase almost 6%

- Leasing cost for a new narrowbody costs approximately $4,000,000/year.

- The ETS equipped aircraft time benefits are equivalent to almost 6 'free' aircraft every year ($23.2 million/year).

- The ETS equipped aircraft can help max out aircraft utilization rates.

Scenario Fleet Size	Aircraft Lease Cost/Year ($)
100	4000000

All figures in this presentation are USD.

There are currently two aircraft parts manufacturers, Safran Aerospace and WheelTug, who are qualifying their aircraft electric taxi system kits according to Federal Aviation Administration regulations. While there has been a request for information along with letters of intent from over 50 major airlines, neither company will give a pricing structure until an airline asks for a quote. To date, no airline has requested a quote even though these electric taxi system kits are expected to be qualified and ready for the airline market in 2022.

As the preceding chapters have demonstrated, there is much to recommend the electronic taxi system for widespread adoption in the aviation industry. These reasons include the fact that commercial airplanes are getting more expensive to operate for several reasons, including the costs for airport expansions, rising wages of airline employees, and the cost of jet fuel, which are all passed on to the airlines and then to paying customers. There is a corporate demand to cut costs and a global demand to reduce emissions. These are just a few of the current issues creating the demand for more fuel-efficient aircraft operations.

What follows is a summary of why the electric taxi system should be considered for installation in narrow-body aircraft:

The cost of jet fuel is the second largest expense to operate an airline. When airlines face near-bankruptcies they react by reducing flights, equipment, personnel, and services. Airline travel is also impacted by economic downturns and security threats or incidents.

Ground support equipment is expensive whether purchased new or used. Many of the replacement parts needed for ground support equipment are specialized and can only be procured from the equipment manufacturer. This monopoly on maintenance parts makes the airline ground support equipment expensive to own.

The electric taxi system components can be electronically configured to work as a regenerative braking system as seen on most modern day electric golf carts. A regenerative brake is a mechanism in which the electric motor that drives the vehicle operates in reverse during braking. This reduces the wear on aircraft break and brake related components.

Aircraft equipped with the electric taxi system will give airport designers the ability to utilize surrounding areas that in the past could not accommodate jet aircraft using main engine thrust for taxi power. As long as there is enough clearance for the aircraft wings and a smooth ground surface below, an aircraft equipped with the electric taxi system could taxi anywhere.

The electric taxi system will give the aircraft better mobility on the ground and better access to airport facilities. Re-design of terminal buildings or re-design of the ramp area can be costly, so ramp redesign is usually not an option. It is in the best interest for a terminal operator or airport manager to use the gates available to the best possible advantage.

Damage caused by foreign object debris at airports can cost the airport tenants and airlines millions of dollars every year. According to The Boeing Aircraft Company, foreign object debris can be anything that does not have a purpose in or near an airplane and as a result cause damage to airplanes and injure airline or airport personnel. Foreign object debris can be caused by loose hardware, building materials, sand, rocks, dead animals, and pieces of luggage. Foreign object debris is found at taxiways, cargo aprons, terminal gates, aircraft run-up areas, and runways. When an aircraft is operated under electrical power with the main engines shut down, the risk of engine damage from foreign object debris is eliminated.

There is a strong future for the single aisle narrow-body aircraft but there are many measures that can affect the decision to promote the design of a new aircraft. Industry experts focus on whether or not to take on the high costs of development in changing the structure of the airplane model. Once the aircraft manufacturer has certified a new model of aircraft for scheduled commercial airline service, it can take them up to and maybe exceeding twenty years to recover their initial investment. There is a strong future for family of Airbus A-320 aircraft and the Boeing B-737. Both of these aircraft are the target airplanes for the first aircraft electric taxi system installation.

The landing gear currently manufactured for the narrow-body aircraft are strong enough to support the new electric taxi system kit as offered from the systems' manufacturers. Landing gear compatibility to support the aircraft electric taxi system installations on smaller and larger aircraft will need to be evaluated. Once proven as a value added asset, the landing gear manufacturers can design landing gear to allow the electric taxi system to be installed during production.

Aircraft equipped with the electric taxi system will contribute less to local airport area air pollution. This contribution should not be underestimated as continued public awareness and growth in air traffic has made the environment a main focus in the future of the commercial aviation industry. It is widely known that sustaining the long-term growth of air transport depends upon environmental improvements for future acceptability.

Airport Benefit: Max Throughput

- The benefits of the ETS are important
- Demand continues to increase
- Expansion slow, expensive (LHR, LGA, MUC, ALY)
- This requires maximization of throughput for all assets: terminals, runways, & aircraft
  ideally without increasing manpower
- Airports that operate more efficiently will attract more traffic (both as a hub and a destination)

More information about the Aircraft Electric Taxi System can be found at:

www. Aircraftelectrictaxisystem.com.

ProQuest Dissertations And Theses; Thesis (D.B.A.) Northcentral University, 2017 Publication Number: AAT 10637574; ISBN: 9780355409222; Source: Dissertation Abstracts International, Volume: 79-02(E), Section: B.; 138 p.

IEEE Trade Journal Publication Number: AAT 10637574; ISBN: 9780355409222

REFERENCES

Adrangi, B, Gritta, R. D., & Raffiee, K. (2014). Dynamic interdependence in jet fuel prices and air carrier revenues. *Atlantic Economic Journal*, 42(4), 473-474. doi.org.proxy1.ncu.edu/10.1007/s11293-013-9388-9

Allignol, C., Barnier, N., Flener, P., & Pearson, J. (2012). Constraint programming for air traffic management: A survey. *The Knowledge Engineering Review*, 27(3), 361-392. doi:.org/10.1017/S0269888912000215

Andrews, M. J., Bergeron, J. H., Creyf, L., Erkelens, C., Halloway, L. B., Murphy, G. F, & Schmitt, D. (2011). International transportation law. *The International Lawyer*, 45(1), 313-327. http://studentorgs.law.smu. edu/International-Law-Review-Association/Journals/ TIL.aspx

Anyan, F. (2013). The influence of power shifts in data collection and analysis stages: A focus on qualitative research interview. *The Qualitative Report*, 18(18), 1-9. http://search.proquest.com.proxy1.ncu.edu/ docview/1505321395?accountid=28180

Assaf, A. G., & Josiassen, A. (2011). The operational performance of UK airlines: 2002-2007. *Journal of Economic Studies*, 38(1), 5-16. doi.org/10.1108/01443581111096114

Atkin, J. A., Burke, E. K., & Greenwood, J. S. (2011). A comparison of two methods for reducing take-off delay at London Heathrow Airport. *Journal of Scheduling*, 14(5), 409-421.doi.org/10.1007/s10951-011-0228-y

Bachtel, B. (2008). Foreign object debris and damage prevention. *Boeing Aero Magazine*. Retrieved 2008-10-28.

Bao, D., Hua, S., & Gu, J. (2016) Relevance of airport accessibility and airport competition *Journal of Air Transport Management*, 55, 52–60

Baxter, P., & Jack, S. (2008). Qualitative case study methodology: Study design and implementation for novice researchers. *The Qualitative Report*, 13(4), 44-559. http://nsuworks.nova.edu/tqr/vol13/iss4/2

Bazargan, M., Lange, D., Tran, L., & Zhou, Z. (2013). A simulation approach to airline cost benefit analysis. *Journal of Management Policy and Practice*, 14(2), 54-61. http://aripd.org/jmp

Bennell, J. A., Mesgarpour, M., & Potts, C. N. (2013). Airport runway scheduling. Annals of *Operations Research*, 204(1), 249-270. doi.org/10.1007/s10479-012-1268-1

Bubalo, B., & Daduna, J. R. (2011). Airport capacity and demand calculations by simulation--the case of Berlin-Brandenburg international Airport. *Netnomics Economic Research and Electronic Networking*, 12(3), 161-181. doi.org/10.1007/s11066-011-9065-6

Burchell, R. W., Crosby, M. S., & Russo, M. (2011). Infrastructure need in the united states, 2010-2030: What is the level of need? how will it be paid for? *The Urban Lawyer*, 42/43(4), 41-66. http://search.proquest.com.proxy1.ncu.edu/docview/868179064?accountid=2818

Cahill, J., Redmond, P., Yous, S., Lacey, G., & Butler, W. (2013). The design of a collision avoidance system for use by pilots operating on the airport ramp and in taxiway areas. *Cognition, Technology & Work*, 15(2), 219-238. doi.org/10.1007/s10111-012-0240-9

Cellini, R., & Kee, J. (2010). Cost effectiveness and cost benefit analysis, chapter 25 of *Handbook of Practical Program Evaluation*, Third Edition, San Francisco, CA: Josey-Bass, 2010.

Charles, D. L., Gaines, L. T., Christensen, D. J., & Dempster, J. B. (2013). U.S. Patent No. 8,485,466. Washington, DC: U.S. Patent and Trademark Office.

Chillson, C. (2014). U.S. Patent No. 3,762,670. Washington, DC: U.S. Patent and Trademark Office.

Creswell, J. (2012). Qualitative inquiry and research design; choosing among five approaches (3d ed.). *Reference and Research Book News*, 27(3) http://search.proquest.com.proxy1.ncu.edu/docview/1018410427?accountid=2818

Chapman, J., & Devolpi, K. (2013). Three reasons for safe GSE operation. Flight Safety Foundation. Retrieved from http://flightsafety.org/

Connelly, L. M. (2016). Trustworthiness in qualitative research. *Medsurg Nursing*, 25(6), 435-436. Retrieved from http://search.proquest.com.proxy1.ncu.edu/docview/1849700459?accountid=28180

Cox, S., Garside, J., Kotsialos, A., & Vitanov, V. (2013). Concise process improvement definition with case studies. *The International Journal of Quality & Reliability Management*, 30(9), 970-990. http://search.proquest.com.proxy1.ncu.edu/docview/1432216346?accountid=28180

Doreswamy, G. R., Kothari, A. S., & Tirumalachetty, S. (2015). Simulating the flavors of revenue management for airlines. *Journal of Revenue and Pricing Management*, 14(6), 421-432.doi.org.proxy1.ncu.edu/10.1057/rpm.2015.42

Dorndorf, U., Jaehn, F., & Pesch, E. (2012). Flight gate scheduling with respect to a reference schedule. *Annals of Operations Research*, 194(1), 177-187. doi.org/10.1007/s10479-010-0809-8

Edwards, G., & Endres, G. (2012). Jane's airline recognition guide. New York, NY: Harper Collins.

Ergas, H. (2009). In defense of cost-benefit analysis. *Agenda: A Journal of Policy Analysis and Reform*, 16(3), 31-40. http://press.anu.edu.au/titles/agenda-a-journal-of-policy-analysis-and-reform-

Espino, N. V. (2013). Prediction of foreign object Debris/Damage (FOD) type for elimination in the aeronautics manufacturing environment through logistic regression model. http://search.proquest.com.proxy1.ncu.edu/docview/1498522734?accountid=28180

Evripidou, L. (2012). M&As in the airline industry: Motives and systematic risk. *International Journal of Organizational Analysis*, 20(4), 435-446. doi. org/10.1108/19348831211268625

Frels, R. K., & Onwuegbuzie, A. J. (2013). Administering quantitative instruments with qualitative interviews: A mixed research approach. *Journal of Counseling and Development: JCD*, 91(2), 184-194. http://Jpurnal+of+Counseling+and+Development&hl=cn&as_=sdt=o&as_vis=1&oi=scolart

Friedrich, C., & Robertson, P. A. (2015). Hybrid-electric propulsion for aircraft. *Journal of Aircraft*, 52(1), 176-189. doi.org/10.2514/1.C032660

Fusch, P. I., & Ness, L. R. (2015). Are we there yet? data saturation in qualitative research. *The Qualitative Report*, 20(9), 1408-1416. http://search.proquest.com.proxy1.ncu. edu/docview/1721368991?accountid=28180

Gao, M., Hong, C. Xu, B., & Ding, R. (2012). Flight rescheduling responding to large-area flight delays. *Kybernetes*, 41(10), 1483-1496. doi. org/10.1108/03684921211276693

Garfors, G. (2014). Tarmac delays and airline passengers. http://www.garfors.com/2014/06/100000-flights-day.html

Gentles, S. J., Charles, C., Ploeg, J. & McKibbon, K. A. (2015). Sampling in qualitative research: Insights from an overview of the methods literature. *The Qualitative Report*, 20(11), 1772-1789. http://search.proquest.com.proxy1.ncu.edu/docview/1750038029?accountid=28180

Goguen, D. (2015). Tarmac delays and airline passenger rights. Retrieved from htpp://www.nolo.com/

legal-encyclopedia/tarmac-delays-airline-passenger-rights-33011.html

Goldberg, A. E., & Allen, K. R. (2015), Communicating qualitative research: Some Practical Guideposts for Scholars. *Journal of Marriage and Family*, 77: 3–22. doi: 10.1111/jomf.12153

Gorjidooz, J., & Vasigh, B. (2010). Aircraft valuation in dynamic air transport industry. *Journal of Business & Economics Research*, 8 (7), 1-16. http://www.journals.cluteonline.com/index.php/JBER

Griffin, A. (2013). Comprehensive examination: Global aviation airways and airport infrastructure. http://search.proquest.com.proxy1.ncu.edu/docview/1554346790?accountid=28180

Guerra, M., Gomes, A. d. O., & Filho, A. I. d. S. (2015). Case study in public administration: A critical review of Brazilian scientific production. *Revista De Contemporânea*, 19(2), 270-289. http://search.proquest.com.proxy1.ncu.edu/docview/1663356146?accountid=28180

Gubisch, M. (2012). In focus: Manufacturers aim for electric taxi ETS by 2016. Flight Global. Retrieved from http://www.flightglobal.com/news/

Gudfinnsson, K., Strand, M., & Berndtsson, M. (2015). Analyzing business intelligence maturity. *Journal of Decision Systems*, 24(1), 37-54. http://search.proquest.com.proxy1.ncu.edu/docview/1658876965?accountid=28180

Guo, R., Zhang, Y. Wang, Q. (2014). Comparison of emerging ground propulsion systems for electrified aircraft taxi operations. *Transportation Research: Part C* July 2014; 44:98-109.

Guy, N. (2013). Electric taxi puts on a show at Paris. Aviation week. Retrieved from http://aviationweek.com/ commercial-aviation/electric-taxi-puts-show-paris

Ha, H., Yoshida, Y., & Zhang, A. (2010). Comparative analysis of efficiency for major northeast Asia airports. *Transportation Journal*, 49(4), 9-23. http://www.astl. org/i4a/pages/index.cfm?pageid=3288

Hing, C. B., Smith, T. O., Hooper, L., Song, F., & Donell, S. T. (2011). A review of how to conduct a surgical survey using a questionnaire. *The Knee*, 18(4), 209-13. doi. org/10.1016/j.knee.2010.10.003

Hoque, Z., Covaleski, M. A., & Gooneratne, T. N. (2013). Theoretical triangulation and pluralism in research methods in organizational and accounting research. *Accounting, Auditing & Accountability Journal*, 26(7), 1170-1198..doi.org/10.1108/AAAJ-May-2012-01024

Hospodka, J. (2014). Cost benefit analysis of electric taxi system aircraft. *Journal of Air Transport Management*, 39 (10), 81-88. http://www.journals.elsevier.com/journal-of-air-transport-management/

Huettinger, M. (2014). What determines the business activities in the airline industry? A theoretical framework. *Baltic Journal of Management*, 9(1), 71-90. doi.org/10.1108/ BJM-04-2013-0053

International Aviation Transport Association (IATA), (2013). Fuel prices remain worryingly high when measured against historical averages http:// centreforaviation.

com/news/iatafuel-prices-remain-worryingly-high-when-measured-against-historical-averages

Jin-Woo, P., & Se-Yeon, J. (2011). Transfer passengers' perceptions of airport service quality: A case study of Inchon International Airport. *International Business Research*, 4(3), 75-82. http://www.ibrusa.com/

Khardi, S., Kurniawan, J. S., Katili, I., & Moersidik, S. (2013). Artificial neural network modeling of healthy risk level induced by aircraft pollutant impacts around Soekarno Hatta International Airport. *Journal of Environmental Protection*, 4(8), 28-39. http://www.scirp.org/journal/jep/

Klemm, A. (2010). Causes, benefits, and risks of business tax incentives. *International Tax and Public Finance*, 17(3), 315-336. doi: 10.1007/s10797-010-9135

Kendirli, S., & Kaya, A. (2016). The evaluation of working capital in airline companies which proceed in bist. *Journal of Economic Development, Environment and People*, 5(1), 39-51. http://search.proquest.com.proxy1.ncu.edu/docview/1812291937?accountid=28180

Ko, Y., & Hwang, H. (2011). Management strategy of full-service carrier and its subsidiary low-cost carrier. *International Journal of Advanced Manufacturing Technology*, 52, 391–405. doi: 10.1007/s00170-010-2726-z

Kuo, W., & Kung, S. (2013). Study of the arrival scheduling simulation for the terminal control area at sung-shun airport. *International Journal of Organizational*

Innovation (Online), 5(3), 192-205. http://www.iaoiusa.org/

Landry, S. J., Farley, T., Hoang, T., & Stein, B. (2012). A distributed scheduler for air traffic flow management. *Journal of Scheduling*, 15(5), 537-551. doi.org/10.1007/s10951-012-0271-3

Lamb, A., Daim, T. U., & Anderson, T. R. (2010). Forecasting airplane technologies. Foresight: *The Journal of Futures Studies, Strategic Thinking and Policy*, 12(6), 38-54. doi.org/10.1108/14636681011089970

Langston, L. S. (2013). Gears galore! *Mechanical Engineering*, 135(4), 51-51, 54. Retrieved from https://www.asme.org/about-asme/news-media/newsletters

Langston, L. S. (2015). Forward future. *Mechanical Engineering*, 137(6), 32-37. http://search.proquest.com.proxy1.ncu.edu/docview/1684436927?accountid=28180.

Lau, H., Nakandala, D., Samaranayake, P., & Shum, P. K. (2016). BPM for supporting customer relationship and profit decision. *Business Process Management Journal*, 22(1), 231-255. http://search.proquest.com.proxy1.ncu.edu/docview/1757877841?accountid=28180

Lee, H. H. (2013). An intelligent optimization system for terminal traffic management http://search.proquest.com.proxy.ncu.edu/docview/1496773891?

Lewandowski, R., Koz Å, O., Krystyna, Szpakowska, M., Gorzata, & Trafny, E. A. (2013). Evaluation of applicability of the sartorius airport MD8 sampler for detection of bacillus endospores in indoor air.

Environmental Monitoring and Assessment, 185(4), 3517-26. doi.org/10.1007/s10661-012-2807-6

Liu, Y., Hanson, M., Gupta, G., Malik W., & Jung Y. (2014). Predictability impacts of airport surface automation. *Transportation Research*, part C44, 128-145. http:/Elsevier.com/locate/trc

Lorell, B. H., Mikita, J. S., Anderson, A., Hallinan, Z. P., & Forrest, A. (2015). Informed consent in clinical research: Consensus recommendations for reform identified by an expert interview panel. *Clinical Trials*, 12(6), 692-695. doi.org/10.1177/1740774515594362

Mambo, D. A., Eftekhari, M. M., Steffen, T., & Ahmad, M. W. (2015). Designing an occupancy flow-based controller for airport terminals. *Building Services Engineering Research & Technology*, 36(1), 51-66. .doi.org/10.1177/0143624414540292

Manohar, B., & Kumar, C. H. V. (2012). Green transportation - public sector transport system. *Asia Pacific Journal of Management & Entrepreneurship Research*, 1(3), 94-107. http://lebanonfoundation.org.in/html/apjmer.html

Manoharan, S. (2013). Innovative double bypass engine for increased performance. http://search.proquest.com.proxy1.ncu.edu/docview/1356706919?accountid=28180.

Marshall, B., Cardon, P., Poddar, A., & Fontenot, R. (2013). Does sample size mater in qualitative research; A review of qualitative interviews in IS research. *The Journal of Computer Information Systems*, 54(1), 11-

22.   http://search.proquest.com.proxy1.ncu.edu/ docview/1471047612?accountid=28180

Marín, Á. G. (2006). Airport management: Taxi planning. *Annals of Operations Research*, 143(1), 191-202. doi. org/10.1007/s10479-006-7381-2

Martinez, V. (2013). Time value of money made simple: A graphic teaching method. *Journal of Financial Education*, 39(1), 96-117. http://www.jfedweb.org/

Mass, C. (2012). Now casting: The promise of new technologies of communication, modeling, and observation. *Bulletin of the American Meteorological Society*, 93(6), 797-809. Retrieved from http://search.proquest.com.proxy1. ncu.edu/docview/1026634422?accountid=28180

Mavin, T. J., & Roth, W. (2015). Optimizing a workplace learning pattern: A case study from aviation. *Journal of Workplace Learning*, 27(2), 112. http://search.proquest.com.proxy1.ncu.edu/ docview/1655240779?accountid=28180

Mazerolle, S., & Goodman, A. (2013). Fulfillment of work-life balance from the organizational perspective: A case study. *Journal of Athletic Training*, 48(5), 668-77. Retrieved from http://search.proquest.com.proxy1. ncu.edu/docview/1445002166?accountid=28180

McHendry, G. (2015). The re (d) active force of the transportation security administration. *Criticism*, 57(2), 211-233. http://search.proquest.com.proxy1. ncu.edu/docview/1767583350?accountid=28180

Medina, D. (2013). Federal aviation administration publication. *Federal Register* Vol.78, No. 241, Monday, December 16, 2013 /Rules and Regulations

Morbey, J., Landman M., & Colin M. (2012). U.S. Patent No. 8,121,786, Washington DC: Patent and Trademark Office.

Murner, R. (2012). Future fuel-efficient gas turbine jet engines run hotter make component testing more critical. *Mechanical Engineering*, 134(2), 50. https://www. asme.org/about-asme/news-media/newsletters

Nikoleris, T., Gupta, G., Kistler, M (2014) Detailed estimation of fuel consumption and emissions during aircraft taxi operations at Dallas/Fort Worth International Airport. *Transportation Research Part D: Transport and Environment*, 16(4), 302–308.

Norin, A., Yuan, D., Granberg, T. A., & Värbrand, P. (2012). Scheduling de-icing vehicles within airport logistics: A heuristic algorithm and performance evaluation. *The Journal of the Operational Research Society*, 63(8), 1116-1125. doi .org/10.1057/jors.2011.100

Orhan, G., & Gerede, E. (2013). A study of the strategic responses of Turkish airline companies to the deregulation in Turkey. *Journal of Management Research*, 5(4), 34-63. http://search.proquest.com.proxy1.ncu. edu/docview/1532533385?accountid=28180

Orr, J. S. (2013). High efficiency thrust vector control allocation. Retrieved from http://search.proquest.com.proxy1. ncu.edu/docview/1367583854?accountid=28180

Peredaryenko, M. S., & Krauss, S. E. (2013). Calibrating the human instrument: Understanding the interviewing experience of novice qualitative researchers. *The Qualitative Report*, 18(43), 1-17. Retrieved from http://search.proquest.com.proxy1.ncu.edu/docview/1505320978?accountid=28180

Pereira, C.F., & dos Reis, F.L. (2011). Regular airlines flying towards a low cost strategy. *International Business Research*, 4(1), 93-99. http://www.ibrusa.com/

Potera, C. (2011). Gate waits for better air. Environmental Health Perspectives, 119(7), A288. http://ehp.niehs.nih.gov/journal-archive/

Raes (2008). Ground handling simulation. http://www.fzt.haw-hamburg.de/pers/Scholz/arbeiten/TextRaes.pdf.

Reevc, H. F. (2013). ASME role in powered flight. *Mechanical Engineering*, 135(12), 20. https://www.asme.org/about-asme/news-media/newsletters

Rothwell, S. (2012). Easy jet to test electric plane-taxiing gear to reduce fuel burn. Bloomberg. http://www.bloomberg.com/.

Roberts, S. (2013). Boeing's fuel efficient 777x features folding wings. http://www.gizmag.com/boeing-launches-777x/29818/

Salah, K. (2013). Environmental impact reduction of commercial aircraft around airports. Less noise and less fuel consumption. *European Transport Research Review*, 6(1), 71-84.doi.org/10.1007/s12544-013-0106-0

Sarma, S. K. (2015). Qualitative research: Examining the misconceptions. *South Asian Journal of Management*,

22(3), 176-191. http://search.proquest.com.proxy1. ncu.edu/docview/1732041530?accountid=28180

Samaranayake, P., & Kiridena, S. (2012). Aircraft maintenance planning and scheduling: An integrated framework. *Journal of Quality in Maintenance Engineering*, 18(4), 432-453. doi.org/10.1108/13552511211281598

Shao, X. Qi, M., & Gao, M. (2012). A risk analysis model of flight operations based on region partition. *Kybernetes*, 41(10), 1497-1508. doi.org/10.1108/03684921211276701

Singh, V., & Sharma, S. K. (2015). Fuel consumption optimization in air transport: A review, classification, critique, simple meta-analysis, and future research implications. *European Transport Research Review*, 7(2), 1-24. doi.org.proxy1.ncu.edu/10.1007/s12544-015-0160-x

Singh, A. K., & Sushil. (2013). Modeling enablers of TQM to improve airline performance. *International Journal of Productivity and Performance Management*, 62(3), 250-275. doi.org/10.1108/17410401311309177

Suen, L. W., Huang, H., & Lee, H. (2014). A comparison of convenience sampling and purposive sampling. *Hu Li Za Zhi*, 61(3), 105-11. http://search.proquest.com.proxy1. ncu.edu/docview/1537381331?accountid=28180

Transportation Research Board. (2015). Methodology to improve quantification of aircraft taxi/idle emissions. http://apps.trb.org/cmsfeed/TRBNetProjectDisplay. asp?ProjectID=3440

VALE (2015). Voluntary airport low emissions program. http://www.faa.gov/airports/environmental/vale/media/valebrochure2.pdf

Vinod, B. (2010). The complexities and challenges of the airline fare management process and alignment with revenue management. *Journal of Revenue and Pricing Management*, 9(12), 137-151. doi: 10.1057/rpm.2008.43

Welch, M. (2014). Exploring the impact of communication technologies on business air travel. *Journal of Organizational Culture, Communication and Conflict*, 18(1), 187-213. http://search.proquest.com.proxy1.ncu.edu/docview/1647822716?accountid=28180

Weld, C., Duarte, M., & Kincaid, R. (2010). A runway configuration management model with marginally decreasing transition capacities. Advances in Operations Research, Retrieved from http://www.hindawi.com/journals/aor/

Wilfred S. (2011). Factors affecting airline profits: Evidence from the Philippines. *Journal of Applied Business Research*, 27(6), 17-22. http://www.cluteinstitute.com/journals/journal-of-applied-business-research-jabr/

Wu, H., Chen, T., & Huang, W. (2012). Adaptive fuzzy control of aircraft ABS based on runway identification. *Sensors & Transducers*, 16, 233-242. http://www.sensorsportal.com/HTML/DIGEST/New_Digest.htm.

Wu, W., & Ying-Kai Liao, (2014). A balanced scorecard envelopment approach to assess airlines' performance.

Industrial Management & Data Systems, 114(1), 123-143. doi.org/10.1108/IMDS-03-2013-0135

Zamfir, M., Manea, M. D., & Ionescu, L. (2016). Return on investment - indicator for measuring the profitability of invested capital. *Valahian Journal of Economic Studies*, 7(2), 1-8. doi:http://dx.doi.org.proxy1.ncu.edu/10.1515/vjes-2016-0010

Zohrabi, M. (2013). Mixed method research: Instruments, validity, reliability and reporting findings. *Theory and Practice in Language Studies*, 3(2), 254-262. http://academypublisher.com/tpls/

APPENDIX A

American Airlines - United Airlines Partial Fleet

Data recorded from Aviation *Week, 2016. Airline Operations Review*

Fleet Data 389 A-320 aircraft. 336 B-737 aircraft

a. Annual fuel purchased by American Airlines 2016. 5,678,808 gallons.

   http://www.transtats.bts.gov/fuel.asp?pn=1

b. Average price per gallon. $1.42 (06.16.2016)

   http://aviationweek.com/%5Bprimary-term%5D/jet-and-avgas-fuel-prices-august-2016

c. 2016 Fleet engine hour's B-737 Family 75,673 A320 Family 29,766

d. 2016 Number of aircraft cycles B-737 Family 15,410 A320 Family 15,227 United Airlines.

Data recorded from Aviation *Week, 2016. Airline Operations Review*

Fleet Data 155 A-320 aircraft. 319 B-737 aircraft

a. Annual jet fuel purchased by United Airlines in 2016 3,215,521 gallons.

   http://www.transtats.bts.gov/fuel.asp?pn=1

b. Average price per gallon. $1.42 (06.16.2016)

> *http://aviationweek.com/%5Bprimary-*
> *term%5D/jet-and-avgas-fuel-prices-august-2016*

c. 2016 Fleet engine hour's B-737 Family <u>1,027,977</u> A320 Family <u>414,779.</u>

d. 2016 Fleet Number of aircraft cycles B-737 Family <u>372,071</u> A320 Family <u>197,936.</u>

The following fuel burn during taxi is typical for Boeing 737 and Airbus A320 family of aircraft.

e. Amount of fuel consumed for 54 minutes at 7% power <u>232 gal.</u>

f. *Fuel Burn CFM-56 engine 850 lb. hr. 14 lb. min. X 54 min. = 756 lbs.*

g. *6.5 lbs. per gallon / 756 lbs. = 116 gal. X 2 engines = 232 gal.*

> *http://aviation.stackexchange.com/*
> *questions/8429/does-the-apu-consume-a-lot-of-fuel-*
> *compared-to-a-jet-*

h. Amount of fuel consumed during each taxi. <u>232 gal.</u>

i. Cost of fuel consumed during taxi. <u>$329.44</u> ($1.42 X 232= $329.44)

ABOUT THE AUTHOR

Johnson holds a Doctor of Business Administration, specializing in Management of Engineering and Technology. He has written articles for the *IEEE* trade journal about the aircraft Electric Green Taxi System (EGTS), and his doctoral dissertation investigated the advantages of the Electric Taxi  System (ETS). While attending Embry Riddle Aeronautical University, he wrote his master's thesis titled: "A Study of the Federal Aviation Administration Latest Ruling on Extended Range Operation with Two Engine Aircraft 'ETOPS [Extended-range Twin-engine Operational Performance Standards].'" This thesis was sent to the Federal Aviation Administration as an answer to a notice of proposed rulemaking (NPRM) and was instrumental with increasing the ETOPS rule from 220 min. to 300 min. Additionally, Johnson holds an FAA commercial pilots license and FAA mechanics certifications.